INTERNATIONAL SERIES IN
NATURAL PHILOSOPHY
General Editor D. ter Haar

VOLUME 92

LECTURES ON SELECTED TOPICS IN STATISTICAL MECHANICS

Other Titles of Interest

DAVYDOV:	Quantum Mechanics, 2nd edition
HEINE:	Group Theory in Quantum Mechanics: An Introduction to its Present Usage, 2nd edition
LANDAU & LIFSHITZ:	Mechanics, 3rd edition
LANDAU & LIFSHITZ:	Statistical Physics, 2nd edition
LANDAU & LIFSHITZ:	Quantum Mechanics - Non-relativistic Theory, 3rd edition
PATHRIA:	Statistical Mechanics
TAYLOR:	Mechanics: Classical and Quantum

LECTURES ON SELECTED TOPICS IN STATISTICAL MECHANICS

by

D. TER HAAR

University Reader in Theoretical Physics and Fellow of Magdalen College, Oxford

PERGAMON PRESS

OXFORD · NEW YORK · TORONTO · SYDNEY · PARIS · FRANKFURT

U.K.	Pergamon Press Ltd., Headington Hill Hall, Oxford OX3 0BW, England
U.S.A.	Pergamon Press Inc., Maxwell House, Fairview Park, Elmsford, New York 10523, U.S.A.
CANADA	Pergamon of Canada Ltd., 75 The East Mall, Toronto, Ontario, Canada
AUSTRALIA	Pergamon Press (Aust.) Pty. Ltd., 19a Boundary Street, Rushcutters Bay, N.S.W. 2011, Australia
FRANCE	Pergamon Press SARL, 24 rue des Ecoles, 75240 Paris, Cedex 05, France
WEST GERMANY	Pergamon Press GmbH, 6242 Kronberg-Taunus, Pferdstrasse 1, West Germany

---

First edition 1977

Library of Congress Cataloging in Publication Data

Haar, D. ter.
Lectures on selected topics in statistical mechanics

Includes index.
1. Statistical mechanics--Addresses, essays, lectures.
I. Title.
QC174.82.H32 1977   530.1'3   77-8300
ISBN 0-08-017937-1

In order to make this volume available as economically and rapidly as possible the author's typescript has been reproduced in its original form. This method unfortunately has its typographical limitations but it is hoped that they in no way distract the reader.

Printed in Great Britain by William Clowes & Sons, Limited London, Beccles and Colchester

Contents

Preface

The present lecture notes were prepared for a set of lectures which were given at the 1971 Simla Summer School of Statistical Mechanics, organized by the Centre of Advanced Study in Theoretical Physics and Astrophysics of the University of Delhi and the Centre for Postgraduate Studies of the University of Himachal Pradesh. Because of lack of time not all the material presented here was covered in my lectures. I want to stress that some parts lean very heavily on the original papers given in the list of references at the end of the notes.

Because these are lecture notes, the choice of material is highly subjective and I have, for obvious reasons, covered those subjects in which I myself have been or am interested. All the same, I hope that this personal selection may prove to be of interest to other people, especially as it contains some unpublished material as well as some material which is not easily available.

I should like to express my thanks to the organizers of the Summer School, and especially to Professor F.C. Auluck for inviting me to give these lectures and for their generous hospitality.

Simla, Himachal Pradesh,
June 1971

D. ter Haar

CHAPTER 1

The Occupation Number Representation

Let us first of all note that although statistical mechanics deals with 'large' systems of interacting particles, whereas field theory deals with particles and quantized fields or with interacting fields, in the second quantization formalism these problems become very similar. This has led to the introduction of diagram techniques in statistical mechanics both at zero and at non-zero temperatures.

Let us also note the need to use representative ensembles because of our restricted knowledge about the actual physical systems we are dealing with. Bearing in mind that the total number of particles in the system is also not known accurately, we are led to the use of grand ensembles. We shall start the introduction with a recapitulation of some relevant formulae from ensemble theory.

1.1 Classical Petit Ensembles

If we consider an ensemble of η systems consisting of N identical atoms (or molecules), if $p_1^{(i)},\ldots,p_s^{(i)}$, $q_1^{(i)},\ldots,q_s^{(i)}$ are the generalized momenta and coordinates of the ith atom, if $D(q_1^{(1)},\ldots,q_s^{(N)},\ p_1^{(1)},\ldots,p_s^{(N)};\ t)d\Omega$ is the number of systems in the ensemble with a representative point, $q_1^{(1)},\ldots,q_s^{(N)},\ p_1^{(1)},\ldots,p_s^{(N)}$, within a volume element $d\Omega$ of phase space or Γ-space, where

$$d\Omega = \prod_{i=1}^{N}\prod_{j=1}^{s} dp_j^{(i)}\, dq_j^{(i)} ,$$

and if we introduce the ensemble density $\rho = D/\eta$, one can show that when we consider an ensemble representing a system in thermodynamic equilibrium at an absolute temperature T, ρ is given by the equation

$$\rho = e^{\beta(\psi - \varepsilon)} , \tag{1.1}$$

where ψ is a function of β, where

$$\beta = \frac{1}{k_B T} \tag{1.2}$$

(k_B : Boltzmann's constant), and where ε is the energy of the system as function of the p's and q's , that is, the Hamiltonian :

$$\varepsilon = H(p,q). \tag{1.3}$$

From the definition of ρ it follows that it must be normalized

$$\int \rho \, d\Omega = 1 \quad , \tag{1.4}$$

where the integration is over the whole of Γ-space. From (1.1) and (1.4) it follows that ψ must satisfy the equation

$$e^{-\beta\psi} = \int e^{-\beta\epsilon} \, d\Omega \, , \tag{1.5}$$

and one shows by standard methods that ψ is the Helmholtz free energy F , apart possibly from an additive multiple of β^{-1}. To fix this constant and ensure the proper transition from quantal formulae we can write

$$e^{-\beta F} = \frac{1}{N!} \int e^{-\beta\epsilon} \frac{d\Omega}{h^{SN}} \equiv Z_{\Gamma}^{(N)}, \tag{1.6}$$

where h is Planck's constant. The function $Z_{\Gamma}^{(N)}$ is called the partition function.

Once we know ψ (or $Z_{\Gamma}^{(N)}$) we can by differentiation find the average values of various phase functions, for which we have the general formula

$$\overline{g(p,q)} = \int g \, \rho \, d\Omega \, . \tag{1.7}$$

For instance, we have

$$\bar{\epsilon} = \frac{\partial \beta \psi}{\partial \beta} \, , \tag{1.8}$$

which is the Gibbs-Helmholtz equation, and for the pressure P

$$P = - \frac{\partial\psi}{\partial V} \, , \tag{1.9}$$

where V is the volume.

If we are dealing with point particles, so that $s = 3$, the kinetic energy part of ϵ is simply $\sum_i p_i^2/2m$, and the integration over the momenta is straightforward and leads to

$$Z_{\Gamma}^{(N)} = Q_N / v_o^N \quad , \tag{1.10}$$

where

$$v_o = \left(\frac{\beta h^2}{2\pi m}\right)^{\frac{3}{2}} \quad , \tag{1.11}$$

which is essentially the cube of the thermal de Broglie wavelength, that is, the de Broglie wavelength of a particle moving with a kinetic energy equal to $k_B T$, and where Q_N is the so-called configurational partition function,

$$Q_N = \frac{1}{N!} \int e^{-U(r_i)} \prod_i d^3 r_i \tag{1.12}$$

where $U(r_i)$ is the total potential energy of the system.

1.2 Classical Grand Ensembles

We now consider an ensemble, the systems of which do not necessarily all contain N particles. If $D(p_1^{(1)},\ldots,p_s^{(N)}, q_1^{(1)},\ldots,q_s^{(N)}; t)d\Omega$ is the number of systems in the ensemble containing N particles and with their representative points within $d\Omega_N$, and if again $\rho = D/\eta$, we now find that for an ensemble representing a system in thermodynamic equilibrium and able to exchange particles with its surroundings, ρ is given by

$$\rho = \frac{1}{N!}\, e^{-q + \alpha N - \beta\varepsilon} \,, \tag{1.13}$$

where β and ε have the same meaning as in (1.1), where α can be shown to be the partial Helmholtz free energy multiplied by β, and where the grand potential - or q-potential -, which because of the normalization of ρ satisfies the equation

$$e^q = \sum_N \frac{1}{N!}\, e^{\alpha N} \int e^{-\beta\varepsilon}\, d\Omega_N \,, \tag{1.14}$$

can be shown to be equal to βPV in the case of a homogeneous system. We note in passing that we can write (1.14) in the form

$$e^q = \sum_N h^{sN}\, e^{\alpha N}\, Z_\Gamma^{(N)} \,. \tag{1.15}$$

The grand-ensemble averages $\langle g\rangle$ of phase functions are now given by the formula

$$\langle g\rangle = \sum_N \int g\,\rho\, d\Omega_N \,, \tag{1.16}$$

and many of them follow from q through differentiation. For example, we have

$$\langle N\rangle = \frac{\partial q}{\partial\alpha} \,, \tag{1.17}$$

$$\langle\varepsilon\rangle = -\frac{\partial q}{\partial\beta} \,, \tag{1.18}$$

$$P = \frac{1}{\beta}\,\frac{\partial q}{\partial V} \,. \tag{1.19}$$

If we are not dealing with stationary ensembles, such as the ones described by (1.1) and (1.13), the ensemble density ρ satisfies the following equation of motion:

$$\dot\rho = \{H, \rho\} \,, \tag{1.20}$$

where the dot indicates a time derivative and the curly brackets a Poisson bracket:

$$\{A,B\} = \sum_{i,j}\left[\frac{\partial A}{\partial q_j^{(i)}}\,\frac{\partial B}{\partial p_j^{(i)}} - \frac{\partial A}{\partial p_j^{(i)}}\,\frac{\partial B}{\partial q_j^{(i)}}\right] \tag{1.21}$$

1.3 Quantal Petit Ensembles

In quantum statistica mechanics the density matrix takes over the rôle of the ensemble density. Its equation of motion is the quantal counterpart of (1.20):

$$i\hbar \dot{\rho}_{op} = \left[H_{op}, \rho_{op}\right]_- , \tag{1.22}$$

where

$$\left[A_{op}, B_{op}\right]_- = A_{op}B_{op} - B_{op}A_{op} \tag{1.23}$$

is a commutator, and where we indicate operators by the subscript 'op' or by a caret '^'.

Averages of phase functions are double averages in the quantal case: once over all the systems in an ensemble, and once for each system over the state of the system. The density matrix, or statistical operator, is defined in such a way that the average $\bar{g}$ of a phase function g is now given by the equation

$$\bar{g} = \mathrm{Tr}\ \rho_{op} g_{op} . \tag{1.24}$$

A special example of (1.24) is the normalization condition

$$\mathrm{Tr}\ \rho_{op} = 1 . \tag{1.25}$$

For an ensemble representating a system in thermodynamic equilibrium, (1.1) is replaced by

$$\rho_{op} = \exp\left[\beta(\psi - H_{op})\right] , \tag{1.26}$$

where H_{op} is the energy (Hamiltonian) operator, and where β and ψ have the same physical meaning as in (1.1). Equations (1.8) and (1.9) are again valid, but (1.5) must be replaced by

$$e^{-\beta\psi} = \mathrm{Tr}\ \exp(-\beta H_{op}) . \tag{1.27}$$

1.4 Quantal Grand Ensembles

Instead of (1.13), we now have

$$\rho_{op} = \exp(-q + \alpha N_{op} - \beta H_{op}) , \tag{1.28}$$

where q, α, and β have the same meaning as before, and where N_{op} is the number operator. Equation (1.25) again holds, and also (1.24) :

$$\langle g \rangle = \mathrm{Tr}\ \rho_{op} g_{op} , \tag{1.29}$$

but the trace involves summation over N - if we work in a representation in which N is diagonal. Equations (1.17) to (1.19) remains the same, while (1.14) is replaced by

$$e^{q} = \mathrm{Tr}\ \exp(\alpha N_{op} - \beta H_{op}) . \tag{1.30}$$

We must bear in mind that, if we work with systems of bosons (or fermions) the complete orthonormal systems which we use to choose for our representation must be properly symmetrised (or antisymmetrised).

1.5 Occupation-number Representation

It is often convenient to use a formalism which employs creation and annihilation operators. This formalism we shall call the occupation-number representation, although it is widely known as the second-quantisation representation. This term, however, is a misnomer, as nowhere is there a further introduction of the quantum of action; it derives from the fact that in this formalism the wavefunctions become operators. We shall use Dirac's bra and ket notation as it is particularly suitable for our discussion.

For a system of N identical particles one can use as a complete orthonormal set of basic functions the set

$$|i_1, i_2, \ldots, i_N\rangle = \frac{1}{N!} \sum_P \epsilon_P \varphi_{i_1}(P_1)\varphi_{i_2}(P_2) \ldots \varphi_{i_N}(P_N) , \tag{1.31}$$

where the φ_i form a complete orthonormal set of single-particle kets (we do not need to and shall not specify the representation in which we are working), where $P_1, P_2, \ldots, P_N$ is a permutation of the N numbers $1, \ldots, N$, where ϵ_P is 1 for bosons, while for fermions $\epsilon_P = +1$ or -1 according as to whether P is an even or odd permutation. The complete orthonormal ket set $|i_1, \ldots, i_N\rangle$ is a set with the proper symmetry features. The bra set corresponding to the ket set (1.31) is

$$\langle i_1, i_2, \ldots, i_N| = \frac{1}{N!} \sum_P \epsilon_P \varphi^*_{i_1}(P_1)\varphi^*_{i_2}(P_2) \ldots \varphi^*_{i_N}(P_N) , \tag{1.32}$$

where φ^*_i is the conjugate complex of φ_i - or the Hermitean conjugate, if we are dealing with particles described by spinors. The sets (1.31) and (1.32) satisfy the orthonormality relation

$$\langle i'_1, i'_2, \ldots, i'_N|i_1, i_2, \ldots, i_N\rangle = \frac{1}{N!} \sum_P \epsilon_P \delta(i_1 - i'_{P_1}) \ldots \delta(i_N - i'_{P_N}) \tag{1.33}$$

as well as a completeness or closure relation. We have assumed here that i is a continuous parameter. If it is a discrete one, the Dirac δ-function must be replaced by a Kronecker δ-function.

If $|\Psi\rangle$ is a wavefunction of the N-particle system with the correct symmetry properties, we have

$$|\Psi\rangle = \int \ldots \int di_1, \ldots, di_N |i_1, \ldots, i_N\rangle\langle i_1, \ldots, i_N|\Psi\rangle \tag{1.34}$$

Similarly, an operator Ω_{op} operating on N-particle functions in the Hilbert space spanned by the set (1.31) will have the form

$$\Omega_{op} = \int, \ldots, \int di_1, \ldots, di_N \; d_1', \ldots, di_N' |i_1, \ldots, i_N\rangle \langle i_1, \ldots, i_N| \Omega_{op} |i_1', \ldots, i_N'\rangle\langle i_1', \ldots, i_N'| \,. \tag{1.35}$$

Let us now consider systems in which the numbers of particles may have any arbitrary value. That is, let us consider the Hilbert space which is the direct product space of the Hilbert spaces corresponding to $0, 1, \ldots, N, \ldots$ particles. These latter Hilbert spaces are spanned by the sets (1.31) with $N = 0, 1, \ldots, N, \ldots$, respectively, and the complete orthonormal set spanning the product space will be the set

$$|0\rangle, \; |i\rangle, \; |i_1, i_2\rangle, \ldots, \; |i_1, i_2, \ldots, i_N\rangle, \ldots, \tag{1.36}$$

where $|0\rangle$ is the vacuum state in which there is no particle present.

We now introduce creation — or construction — operators which will produce a ket corresponding to N particles from one corresponding to $N-1$ particles in the following way

$$\hat{a}^+(i) |i_1, \ldots, i_N\rangle = \sqrt{N+1}\, |i, i_1, \ldots, i_N\rangle \,. \tag{1.37}$$

From (1.37) it follows that all the $|i_1, \ldots, i_N\rangle$ can be obtained from the vacuum state by the repeated action of creation operators:

$$|i_1, \ldots, i_N\rangle = \frac{1}{\sqrt{N!}}\, \hat{a}^+(i_1)\, \hat{a}^+(i_2), \ldots, \hat{a}^+(i_N) |0\rangle \,. \tag{1.38}$$

From the definitions (1.37) and (1.31) one can prove — and we leave the proof to the reader — that the $\hat{a}^+(i)$ satisfy the (anti)commutation relations

$$\left[\hat{a}^+(i), \hat{a}^+(j)\right]_- = 0 \,, \quad \text{for bosons;} \tag{1.39a}$$

$$\left[\hat{a}^+(i), \hat{a}^+(j)\right]_+ = 0 \,, \quad \text{for fermions;} \tag{1.39b}$$

where

$$\left[A_{op}, B_{op}\right]_+ = A_{op}B_{op} + B_{op}A_{op} \tag{1.40}$$

is an anticommutator.

Consider now the equations which are the adjoint of (1.37):

$$\langle i_1, \ldots, i_N| \hat{a}(i) = \sqrt{N+1}\, \langle i, i_1, \ldots, i_N| \,, \tag{1.41}$$

where $\hat{a}(i)$ is the adjoint of $\hat{a}^+(i)$. From (1.41) it follows that

$$\left[\hat{a}(i), \hat{a}(j)\right]_- = 0 \,, \quad \text{for bosons;} \tag{1.42a}$$

$$\left[\hat{a}(i), \hat{a}(j)\right]_+ = 0 \,, \quad \text{for fermions.} \tag{1.42b}$$

To find out what $\hat{a}(i)$ does to a ket vector, we consider the matrix element

$$\langle i_1', \ldots, i_{N-1}' | \hat{a}(i) | i_1, \ldots, i_N \rangle .$$

We have written $N-1$ indices on the left and N indices on the right because (i) equation (1.41) tells us that $\hat{a}(i)$ operating from the right on an N-particle function produces an $(N+1)$-particle function, and (ii) $|i_1, \ldots, i_N\rangle$ is orthogonal onto $|i_1', \ldots, i_{N'}'\rangle$, if $N \neq N'$. From equations (1.41) and (1.31) or (1.33) it follows that

$$\langle i_1', \ldots, i_{N-1}' | \hat{a}(i) | i_1, \ldots, i_N \rangle = \frac{1}{\sqrt{N}} \Big\{ \delta(i - i_1) \langle i_1', \ldots, i_{N-1}' | i_2, \ldots, N \rangle$$

$$\pm \delta(i - i_2) \langle i_1', \ldots, i_{N-1}' | i_1, i_3, \ldots, i_N \rangle + \ldots +$$

$$(\pm 1)^{N-1} \delta(i - i_N) \langle i_1', \ldots, i_{N-1}' | i_1, \ldots, i_{N-1} \rangle \Big\} \tag{1.43}$$

As equation (1.43) must hold for any bra $\langle i_1', \ldots, i_{N-1}' |$, we have

$$\hat{a}(i) | i_1, \ldots, i_N \rangle = \frac{1}{\sqrt{N}} \Big\{ \delta(i - i_1) | i_2, \ldots, i_N \rangle \pm \delta(i - i_2) | i_1, i_3, \ldots, i_N \rangle$$

$$+ \ldots + (\pm 1)^{N-2} \delta(i - i_{N-1}) | i_1, \ldots, i_{N-2}, i_N \rangle$$

$$+ (\pm 1)^{N-1} \delta(i - i_N) | i_1, \ldots, i_{N-1} \rangle . \tag{1.44}$$

In equations (1.43) and (1.44) the upper (lower) signs correspond to the boson (fermion) case, From equation (1.44) we see that the $\hat{a}(i)$ are annihilation operators. Provided the state i is represented in the ket $|i_1, \ldots, i_N\rangle$, $\hat{a}(i)$ operating upon that ket produces another ket where that state has been removed. In the case of bosons the state i may occur more than one, and there may be more than one non-vanishing term on the right-hand side of (1.44).

From equations (1.44), (1.37), and (1.41) we find the (anti)commutation relations

$$\left[\hat{a}(i), \hat{a}^+(j)\right]_- = \delta(i - j), \quad \text{for bosons;} \tag{1.45a}$$

$$\left[a(i), a^+(j)\right]_+ = \delta(i - j), \quad \text{for fermions.} \tag{1.45b}$$

Let us now consider (1.35) for the special case where Ω_{op} is of the form

$$\Omega_{op} = \sum_{i=1}^{N} \hat{\Omega}_i^{(1)} + \frac{1}{2} \sum_{i,j=1}^{N} \hat{\Omega}_{ij}^{(2)} , \tag{1.46}$$

where the $\hat{\Omega}_i^{(1)}$ are single-particle operators and the $\hat{\Omega}_{ij}^{(2)}$ are two-particle operators. Moreover, let us assume that the $\hat{\Omega}_i^{(1)}$ have all the same form and differ only in the particle on which they operate and that, mutatis mutandis, the same is true for the $\hat{\Omega}_{ij}^{(2)}$ (compare equations (1.49) below). From equation (1.35) and equation (1.38) and its adjoint, we find, if we also use the orthonormality and symmetry properties of the kets :

$$\Omega_{op} = \int di\, di' \langle i|\hat{\Omega}^{(1)}|i'\rangle \hat{a}^{+}(i)\, \hat{a}(i')$$
$$+ \frac{1}{2}\int di\, dj\, di'\, dj'\, \langle ij|\hat{\Omega}^{(2)}|i'j'\rangle \hat{a}^{+}(i)\, \hat{a}^{+}(j)\, \hat{a}(j')\, \hat{a}(i') \quad . \tag{1.47}$$

A case of particular interest, as it often is applicable, is the one where we choose the φ_i to be plane waves in a volume V , that is, where we use the momentum eigenfunctions in the coordinate representation :

$$\varphi_i \rightarrow V^{-\frac{1}{2}}\, e^{i(\underline{k}\cdot\underline{r})} \quad , \tag{1.48}$$

where the $\underline{k}$ form a discrete set, if the volume V is finite. In that case we write $\hat{a}_{\underline{k}}^{+}$ and $\hat{a}_{\underline{k}}$ rather than $\hat{a}^{+}(\underline{k})$ and $\hat{a}(\underline{k})$. Let in the coordinate representation the $\hat{\Omega}_i^{(1)}$ and $\hat{\Omega}_{ij}^{(2)}$ have the form

$$\hat{\Omega}_i^{(1)} = -\frac{\hbar^2}{2m}\, \nabla_i^2 + U(r) \quad , \quad \hat{\Omega}_{ij}^{(2)} = v(\underline{r}_i - \underline{r}_j) \quad , \tag{1.49}$$

which corresponds to Ω_{op} being the Hamiltonian of a system of point particles of mass m moving in a common potential $U(\underline{r})$ and interacting solely through binary forces governed by the interaction potential $v(\underline{r}_i - \underline{r}_j)$. We now get from equation (1.47)

$$\Omega_{op} = \sum_{k} \frac{\hbar^2 k^2}{2m}\, \hat{a}_{\underline{k}}^{+}\hat{a}_{\underline{k}} + \frac{1}{V}\sum_{\underline{k},\underline{q}} U(q)\, \hat{a}_{\underline{k}}^{+}\hat{a}_{\underline{k}-\underline{q}} + \frac{1}{2V}\sum_{\underline{k},\underline{k}',\underline{q}} v(\underline{q})\, \hat{a}_{\underline{k}}^{+}\hat{a}_{\underline{k}'}^{+}\hat{a}_{\underline{k}'+\underline{q}}\hat{a}_{\underline{k}-\underline{q}} \tag{1.50}$$

where

$$U(\underline{q}) = \int d^3\underline{r}\, e^{-i(\underline{q}\cdot\underline{r})}\, U(\underline{r}) \quad , \quad v(\underline{q}) = \int d^3 r\, e^{-i(\underline{q}\cdot\underline{r})}\, v(\underline{r}) \quad . \tag{1.51}$$

We note in conclusion that the operators

$$n_{op}(i) = \hat{a}^{+}(i)\, \hat{a}(i) \quad , \tag{1.52}$$

the so-called occupation-number operators, will when operating upon a ket $|i_1, \ldots, i_N\rangle$ reproduce this ket, multiplied by an integer which is equal to the number of times i occurs amongst the numbers $i_1, \ldots, i_N$, as follows from equation (1.44). One can thus also characterize the kets $|i_1, \ldots, i_N\rangle$ by the numbers of particles which are in a given single-particle state:

$$|i_1, \ldots, i_N\rangle \rightarrow |n(j_1)\, , n(j_2)\, , \ldots \rangle \quad , \tag{1.53}$$

where now the $j_1, j_2, \ldots$ are all different. In the case of bosons the $n(j)$ can take on any integral value, but in the case of fermions they must be 0 or 1. We shall not further develop the formalism using the $n(j)$-representation.

CHAPTER 2
The Green Function Method in Statistical Mechanics

2.1 The Double-time Temperature-dependent Green Functions

There are many cases when we want to know the thermodynamic averages of thermodynamic quantities and it is, therefore, of interest to have a formalism which enables us to evaluate such averages – either exactly, or at least in some more or less well-defined approximation. The Green function formalism, which was first introduced by Bogolyubov and Tyablikov, is such a scheme. It has turned out to be especially useful in the theory of magnetism, as we shall see in these lectures.

Let H_{op} be the Hamiltonian of the system we are considering and let N_{op} be the number operator for the particles in the system. If again

$$\beta = \frac{1}{k_B T} , \tag{1.1}$$

(k_B : Boltzmann's constant, T : absolute temperature), we have for the grand ensemble average of any thermodynamic quantity A , the equation

$$\langle A \rangle = \frac{\mathrm{Tr}\, A_{op} \exp\left[-\beta\,(H_{op} - \mu N_{op})\right]}{\mathrm{Tr}\,\exp\left[-\beta\,(H_{op} - \mu N_{op})\right]} , \tag{1.2}$$

where μ is the partial thermodynamic or chemical potential.

Sometimes it is just as convenient to use a canonical ensemble as a grand ensemble. In that case, we have instead of equation (1.2)

$$\langle A \rangle = \frac{\mathrm{Tr}\; A_{op}\;\exp(-\beta H_{op})}{\mathrm{Tr}\;\exp(-\beta H_{op})} . \tag{1.3}$$

We now define in the usual way time-dependent operators. If we are using a canonical ensemble, we use the normal Heisenberg representation :

$$A_{op}(t) = \exp(i\,H_{op}\,t)\;A_{op}\;\exp(-\,i\,H_{op}\,t) \tag{1.4}$$

(we are using units in which $\hbar = 1$). However, if we are using a grand ensemble, it is more convenient to generalize equation (1.4) and to define $A_{op}(t)$ through the equation

$$A_{op}(t) = \exp(i H'_{op} t) \, A_{op} \, \exp(-i H'_{op} t) \, , \qquad (1.5)$$

where

$$H'_{op} = H_{op} - \mu N_{op} \, . \qquad (1.6)$$

We can now define the double-time temperature-dependent Green functions through the equation

$$\langle\langle A_{op}(t) \, ; B_{op}(t') \rangle\rangle_{\substack{a \\ r}} = \pm i \, \theta(\mp t \pm t') \left\{ \langle A_{op}(t) \, B_{op}(t') \rangle - \eta \langle B_{op}(t') \, A_{op}(t) \rangle \right\} \qquad (1.7)$$

where η is a parameter which we shall choose to be either $+1$ or -1, but which we leave undetermined for the moment. The function θ is the Heaviside function

$$\begin{aligned} \theta(t) &= 0 \, , \; t < 0 \; ; \\ &= 1 \, , \; t > 0 \; ; \end{aligned} \qquad (1.8)$$

and the indices 'a' and 'r' indicate the advanced and retarded Green functions. We shall not introduce here the causal Green functions.

We now want the equation of motion for the Green functions. We first of all note that as the same operator H'_{op} occurs in equations (1.2) and (1.5) and the same operator H_{op} in equation (1.3) and equation (1.4), and as the trace of a product of operators is invariant under a cyclic permutation of the operators, the Green functions depend on t and t' only through the combination $t - t'$. Let us denote differentiation with respect to t by a dot. We shall use the relations

$$i \dot{A}_{op} = [A_{op} \, , H'_{op}]_{-} \quad , \qquad (1.9)$$

$$i \dot{A}_{op} = [A_{op} \, , H_{op}]_{-} \quad , \qquad (1.10)$$

according to whether we are dealing with a grand or with a canonical ensemble, and where

$$[a_{op}, b_{op}]_{\mp} = a_{op} b_{op} \mp b_{op} a_{op} \, . \qquad (1.11)$$

Moreover, we have

$$\dot{\theta}(t - t') = - \dot{\theta}(t' - t) = \delta(t - t') \, , \qquad (1.12)$$

where $\delta(t)$ is the Dirac delta-function. We thus find

$$\begin{aligned} i \, \langle\langle A_{op}(t) \, \dot{;} \, B_{op}(t') \rangle\rangle &= \delta(t - t') \left\{ \langle A_{op}(t) \, B_{op}(t') \rangle - \eta \langle B_{op}(t') \, A_{op}(t) \rangle \right\}^{\cdot} \\ &\quad + \langle\langle [A_{op}(t) \, , H'_{op}]_{-} \, ; B_{op}(t') \rangle\rangle \quad , \end{aligned} \qquad (1.13)$$

or

$$i \langle\langle \dot{A}_{op}(t) ; B_{op}(t')\rangle\rangle = \delta(t-t')\{\langle A_{op}(t)\, B_{op}(t')\rangle - \eta \langle B_{op}(t')\, A_{op}(t)\rangle\} + \langle\langle [A_{op}(t), H_{op}]_- ; B_{op}(t')\rangle\rangle . \quad (1.14)$$

We note: (i) that the equation of motion is the same for the retarded as for the advanced Green function, and (ii) that in equations (1.13) and (1.14) more complicated Green functions appear.

We now introduce the Fourier transform of the Green functions by the equations

$$\langle\langle A_{op}(t) ; B_{op}(t')\rangle\rangle = \int_{-\infty}^{+\infty} \langle\langle A_{op};B_{op}\rangle\rangle_E \, e^{-iE(t-t')}\, dE , \quad (1.15)$$

$$\langle\langle A_{op} ; B_{op}\rangle\rangle_E = \frac{1}{2\pi} \int_{-\infty}^{+\infty} \langle\langle A_{op}(t) ; B_{op}(t')\rangle\rangle \, e^{iE(t-t')}\, d(t-t'). \quad (1.16)$$

Dropping the index 'E' and only writing down the equations for the canonical case, knowing that the grand-canonical case can be obtained by changing H_{op} to H'_{op}, we can write equation (1.14) in the form

$$E \langle\langle A_{op} ; B_{op}\rangle\rangle = \frac{1}{2\pi} \langle A_{op}B_{op} - \eta B_{op}A_{op}\rangle + \langle\langle [A_{op},H_{op}]_- ; B_{op}\rangle\rangle . \quad (1.17)$$

We must now discuss spectral representations. We define first of all the correlation functions

$$F_{BA}(t,t') = \langle B_{op}(t')\, A_{op}(t)\rangle , \quad F_{AB}(t,t') = \langle A_{op}(t)\, B_{op}(t')\rangle , \quad (1.18)$$

and note that the Green functions are linear combinations of F_{BA} and F_{AB}. We now define a function $J(\omega)$ by the equation

$$F_{BA}(t,t') = \int_{-\infty}^{+\infty} J(\omega)\, e^{-i\omega(t-t')}\, d\omega . \quad (1.19)$$

From equations (1.18) and (1.3), and using a representation in which H_{op} is diagonal,

$$\langle \nu | H_{op} | \mu\rangle = \delta_{\nu\mu}\, E_\nu , \quad (1.20)$$

we find after some straightforward calculations

$$J(\omega) = \frac{1}{Z} \sum_{\nu,\mu} \langle \nu|B_{op}|\mu\rangle\langle\mu|A_{op}|\nu\rangle\, e^{-\beta E_\nu}\, \delta(\omega - E_\nu + E_\mu) , \quad (1.21)$$

where

$$Z = \mathrm{Tr}\, \exp(-\beta H_{op}) . \quad (1.22)$$

Similarly, if we write

$$F_{AB}(t,t') = \int_{-\infty}^{+\infty} J'(\omega)\, e^{-i\omega(t-t')}\, d\omega , \quad (1.23)$$

we find

$$J'(\omega) = J(\omega)\, e^{\beta\omega} . \quad (1.24)$$

Consider now the retarded (advanced) Green functions and their Fourier transforms and write

$$\theta(t) = e^{-\epsilon t} \ (\epsilon \to 0+) \ , \ t > 0 ; \quad \theta(t) = 0 \ , \ t < 0 \ . \tag{1.25}$$

We then get

$$\langle\langle A_{op} ; B_{op} \rangle\rangle_{\substack{r \\ a}} = \frac{1}{2\pi} \int_{-\infty}^{+\infty} (e^{\beta\omega} - \eta) \, J(\omega) \, \frac{d\omega}{E - \omega \pm i\epsilon} \ . \tag{1.26}$$

The function $G(E)$, defined by

$$G(E) = \frac{1}{2\pi} \int_{-\infty}^{+\infty} (e^{\beta\omega} - \eta) \, J(\omega) \, \frac{d\omega}{E - \omega} \ , \tag{1.27}$$

has been shown by Bogolyubov and Parasyuk to be an analytic function of E equal to $\langle\langle A_{op} ; B_{op} \rangle\rangle_r$ in the upper half-plane and equal to $\langle\langle A_{op} ; B_{op} \rangle\rangle_a$ in the lower half-plane, while it has singularities on the real axis.

From equation (1.27) and the relation

$$\lim_{\epsilon \to 0+} \frac{1}{x \pm i\epsilon} = \mathcal{P} \frac{1}{x} \mp \pi i \, \delta(x) \ , \tag{1.28}$$

where $\mathcal{P}$ indicates that on integrating we must take the principal part of the integral, we find

$$(e^{\beta\omega} - \eta) \, J(\omega) = - i \lim_{\epsilon \to 0+} \left[G(\omega + i\epsilon) - G(\omega - i\epsilon) \right] \ , \tag{1.29}$$

$$J(\omega) = \frac{-i}{e^{\beta\omega} - \eta} \lim_{\epsilon \to 0+} \left[G(\omega + i\epsilon) - G(\omega - i\epsilon) \right] . \tag{1.30}$$

Strictly speaking, in the case where $\eta = +1$, the right-hand side of equation (1.30) may contain a term $C \, \delta(\omega)$; however, as $C = 0$ in many cases, we shall not consider this term.

2.2 Simple Applications

Let us first consider the case of a perfect gas and take the grand-ensemble case with

$$H'_{op} = \sum_{\underline{k}} (\epsilon_{\underline{k}} - \mu) \, \hat{a}^{+}_{\underline{k}} \hat{a}_{\underline{k}} \ , \tag{2.1}$$

where the $\epsilon_{\underline{k}}$ are the single-particle energies, and where $\hat{a}^{+}_{\underline{k}}$ and $\hat{a}_{\underline{k}}$ are the creation and annihilation operators which satisfy the (anti)commutation relations

$$\left[\hat{a}^{+}_{\underline{k}}, \hat{a}^{+}_{\underline{k}'} \right]_{\mp} = \left[\hat{a}_{\underline{k}}, \hat{a}_{\underline{k}'} \right]_{\mp} = 0 \ , \quad \left[\hat{a}_{\underline{k}}, \hat{a}^{+}_{\underline{k}'} \right]_{\mp} = \delta_{\underline{k}\,\underline{k}'} \ , \tag{2.2}$$

where the upper sign refers to bosons and the lower sign to fermions, while $\delta_{\underline{k}\,\underline{k}'}$ is the Kronecker symbol ,

$$\delta_{\underline{k}\underline{k}'} = 0\ ,\ \underline{k} \neq \underline{k}'\ ;\ \delta_{\underline{k}\underline{k}} = 1\ . \tag{2.3}$$

It is convenient to choose $\eta = +1$ for bosons and $\eta = -1$ for fermions. Choosing $A_{op} = \hat{a}_{\underline{k}}$, $B_{op} = \hat{a}^{+}_{\underline{k}}$, we then get from equations (1.17) and (2.1) :

$$E\,\langle\langle \hat{a}_{\underline{k}}\,;\,\hat{a}^{+}_{\underline{k}}\rangle\rangle = \frac{1}{2\pi} + (\epsilon_{\underline{k}} - \mu)\,\langle\langle \hat{a}_{\underline{k}}\,;\,\hat{a}^{+}_{\underline{k}}\rangle\rangle\ ,$$

or

$$\langle\langle \hat{a}_{\underline{k}}\,;\,\hat{a}^{+}_{\underline{k}}\rangle\rangle = \frac{1}{2\pi(E - \epsilon_{\underline{k}} + \mu)} \tag{2.4}$$

From equations (1.30), (1.28), (1.19) and (2.4) we then get

$$\langle a^{+}_{\underline{k}}(t')\ a_{\underline{k}}(t)\rangle = e^{-(\epsilon_{\underline{k}} - \mu)(t-t')}\left[e^{\beta(\epsilon_{\underline{k}} - \mu)} - \eta\right]^{-1}, \tag{2.5}$$

which for $t = t'$ leads to the usual Bose-Einstein and Fermi-Dirac formulae for the average occupation number $\langle n_{\underline{k}}\rangle = \langle a^{+}_{\underline{k}} a_{\underline{k}}\rangle$.

Consider now a system with a Hamiltonian $\hat{H}_o$ perturbed by a periodic term $U_{op}\, e^{i\omega t + \epsilon t}$ $(\epsilon \to 0+)$. Let the density matrix at $t = -\infty$ be the equilibrium density matrix corresponding to $\hat{H}_o$:

$$\rho_{op}(-\infty) = \hat{\rho}_o = \exp(-\beta\hat{H}_o)/Z_o\ ,\quad Z_o = \mathrm{Tr}\ e^{-\beta\hat{H}_o}\ . \tag{2.6}$$

If we write

$$\rho_{op}(t) = \hat{\rho}_o + \Delta\rho_{op}\ ,$$

we get from the equation of motion for the density matrix ,

$$i\,\dot{\rho}_{op} = \left[H_{op}\,,\,\rho_{op}\right]_{-}\ , \tag{2.7}$$

$$i\,\Delta\dot{\rho}_{op} = \left[U_{op}\,,\,\hat{\rho}_o\right]_{-}\ e^{i\omega t + \epsilon t} + \left[\hat{H}_o\,,\,\Delta\rho_{op}\right]_{-}\ , \tag{2.8}$$

where we have neglected a second-order term involving $[U_{op}\,,\Delta\rho_{op}]_{-}$.

Introducing

$$\Delta\hat{\rho}' = e^{i\hat{H}_o t}\ \Delta\rho_{op}\ e^{-i\hat{H}_o t}\ , \tag{2.9}$$

substituting this into equation (2.8), integrating, and again using equation (2.9), we get

$$\Delta\rho_{op} = -i\int_{-\infty}^{t} e^{i\hat{H}_o(\tau-t)}\left[U_{op},\hat{\rho}_o\right]_{-}\ e^{-i\hat{H}_o(\tau-t)}\ e^{i\omega\tau + \epsilon\tau}\,d\tau \tag{2.10}$$

If G_{op} is an operator, corresponding to some physical quantity, we get for the average of this physical quantity

$$\langle \hat{G}(t) \rangle = \text{Tr } \rho_{op} G_{op} = \text{Tr } \hat{\rho}_o G_{op} + \text{Tr } \Delta \rho_{op} G_{op}$$

$$= \langle G \rangle_o - i \int_{-\infty}^{t} \langle [G'_{op}(t) , U'_{op}(\tau)]_- \rangle_o \, e^{i\omega\tau + \epsilon\tau} \, d\tau$$

$$= \langle G \rangle_o + 2\pi \, e^{i\omega t + \epsilon t} \langle\langle G_{op} ; U_{op} \rangle\rangle_{r, \eta = +1} \quad , \tag{2.11}$$

where

$$G'_{op}(t) = e^{i\hat{H}_o t} \, G_{op} \, e^{-i\hat{H}_o t} , \quad U'_{op}(\tau) = e^{i\hat{H}_o \tau} \, U_{op} \, e^{-i\hat{H}_o \tau} . \tag{2.12}$$

where $\langle \dots \rangle_o = \text{Tr} \, (\hat{\rho}_o \dots)$, and where we have used the relation

$$\int_{-\infty}^{t} f(\tau) \, d\tau = \int_{-\infty}^{+\infty} \theta(t - \tau) \; f(\tau) \, d\tau . \tag{2.13}$$

We shall apply equation (2.11), the so-called Kubo formula, to the case of ferromagnetic resonance in subsection 2.4.

To conclude this subsection, we note that for the case of a perfect gas the poles of the Green function $\langle\langle \hat{a}_{\underline{k}} ; \hat{a}^+_{\underline{k}} \rangle\rangle$ occur at the single-particle energies (reckoned from the chemical potential as zero). One might hope that also in the general case the poles of the single-particle Green function $\langle\langle \hat{a}_{\underline{k}} ; \hat{a}^+_{\underline{k}} \rangle\rangle$ would give us information about the energy spectrum of the system. From equations (1.21) and (1.27) it follows that (E' and $\hat{H}'$ are connected by the equation $\langle \nu | H'_{op} | \mu \rangle = \delta_{\nu\mu} E'_\nu$; compare equation (1.20))

$$G(E) = \frac{1}{2\pi Z} \sum_{\nu, \mu} \left[e^{-\beta E'_\nu} \frac{\langle \nu | A_{op} | \mu \rangle \langle \mu | B_{op} | \nu \rangle}{E - E'_\mu + E'_\nu} - \eta \frac{\langle \nu | B_{op} | \mu \rangle \langle \mu | A_{op} | \nu \rangle}{E - E'_\nu + E'_\mu} \right] , \tag{2.14}$$

so that the poles of <u>any</u> Green function give the system energy <u>differences</u> $E'_\nu - E'_\mu$. Introducing the function

$$G(E) = \sum_{\mu} \frac{|\mu\rangle\langle\mu|}{E'_\mu - E} . \tag{2.15}$$

which satisfy the equation

$$(H'_{op} - E) \, g(E) = 1 \quad , \tag{2.16}$$

we can write equation (2.14) in the form

$$G(E) = \frac{1}{2\pi Z} \sum_{\mu} e^{-\beta E'_\mu} \langle \mu | - A_{op} \, g(E'_\mu + E) \, B_{op} - \eta B_{op} \, g(E'_\mu - E) \, A_{op} | \mu \rangle . \tag{2.17}$$

If we indicate by a superscript how many particles there are in the system and assume that not only H_{op}, but also N_{op} is diagonal in the representation in which we are working, we have

$$E^{(N)}_\nu = E'^{(N)}_\nu + \mu N . \tag{2.18}$$

Moreover, for large N we have from the physical meaning of μ (the subscript 'o' indicates the ground state)

$$\mu = E_o^{(N+1)} - E_o^{(N)} \ . \tag{2.19}$$

Consider now the functions

$$G_\nu(E) = \langle \nu | - A_{op}\, g(E'_\nu + E)\, B_{op} - \eta\, B_{op}\, g(E'_\nu - E)\, A_{op} | \nu \rangle \ . \tag{2.20}$$

Since

$$\langle \nu | A_{op} \left\{ (H'_{op} - E - E'_\nu)\, g(E'_\nu + E) - 1 \right\} B_{op} - \eta B_{op} \left\{ (H'_{op} + E - E'_\nu)\, g(E'_\nu - E) A_{op} | \nu \rangle \right\} = 0 \,, \tag{2.21}$$

it follows that the $G_\nu(E)$ satisfy equation (1.17). They have also the same poles as $G(E)$.

For the special case where $\nu = 0$, $\hat{A} = \hat{a}_{\underline{k}}^+$, $\hat{B} = \hat{a}_{\underline{k}}^+$, we have a function $G_o^{(N)}(\hat{a}_{\underline{k}}, \hat{a}_{\underline{k}}^+)$ which is the $T = 0$ limit of $\langle\langle \hat{a}_{\underline{k}} ; \hat{a}_{\underline{k}}^+ \rangle\rangle$, which satisfies equation (1.17) and which according to equation (2.14) can be written as

$$G_o^{(N)}(\hat{a}_{\underline{k}}, \hat{a}_{\underline{k}}^+) = \frac{1}{2\pi} \sum_\mu \left\{ \frac{\langle 0^{(N)} | \hat{a}_{\underline{k}} | \mu^{(N+1)} \rangle \langle \mu^{(N+1)} | \hat{a}_{\underline{k}}^+ | 0^{(N)} \rangle}{E - E'^{(N+1)} + E_o'^{(N)}} - \eta \frac{\langle 0^{(N)} | \hat{a}_{\underline{k}}^+ | \mu^{(N-1)} \rangle \langle \mu^{(N-1)} | \hat{a}_{\underline{k}} | 0^{(N)} \rangle}{E - E_o'^{(N)} + E_\mu'^{(N-1)}} \right\} \tag{2.22}$$

Using equations (2.18) and (2.19), we find that the poles corresponding to the first term occur for

$$E = E_\mu^{(N+1)} - E_o^{(N+1)} \,, \tag{2.23}$$

and those for the second term

$$E = E_\mu^{(N-1)} - E_o^{(N-1)} - \left\{ E_o^{(N+1)} - 2E_o^{(N)} + E_o^{(N-1)} \right\} \ . \tag{2.24}$$

If there is no pairing energy, the term in the braces tends to zero as $N \to \infty$, and we see that the poles are just the excitation energies of the system. If the $\hat{a}_{\underline{k}}$ and $\hat{a}_{\underline{k}}^+$ are approximately the correct (quasi-)particle operators, the excitation energies will be approximately the single-(quasi-)particle energies.

2.3 The Heisenberg Ferromagnet

Consider the spin-$\frac{1}{2}$ case with a Hamiltonian given by

$$H_{op} = - g\mu_B B \sum_{\underline{f}} \hat{S}^z_{\underline{f}} - 2 \sum_{\underline{f},\underline{g}} I(\underline{f}-\underline{g})\ (\hat{S}_{\underline{f}}\cdot\hat{S}_{\underline{g}}) , \tag{3.1}$$

where g is the Landé factor which we shall put equal to 2, μ_B the Bohr magneton, B the strength of the magnetic induction (applied along the z-axis), $\hat{S}_{\underline{f}}$ the spin operator vector for the spin on site $\underline{f}$ which in our case has the components

$$\hat{S}^x = \frac{1}{2}\begin{pmatrix} 0 & 1 \\ 1 & 0 \end{pmatrix}, \quad \hat{S}^y = \frac{1}{2}\begin{pmatrix} 0 & -i \\ i & 0 \end{pmatrix}, \quad \hat{S}^z = \frac{1}{2}\begin{pmatrix} 1 & 0 \\ 0 & -1 \end{pmatrix}, \tag{3.2}$$

as far as site $\underline{f}$ is concerned, while it is the unit operator for the other sites, and $I(\underline{f}-\underline{g})$ are exchange integrals which are functions of $\underline{f}-\underline{g}$, and which we shall assume to satisfy the equations

$$I(0) = 0 , \quad I(\underline{f}-\underline{g}) = I(\underline{g}-\underline{f}) . \tag{3.3}$$

For the spin-$\frac{1}{2}$ case it is convenient to introduce instead of the $\hat{S}_{\underline{f}}$ new operators by the equations

$$\hat{b}_{\underline{f}} = \hat{S}^+_{\underline{f}} = \hat{S}^x_{\underline{f}} + i\hat{S}^y_{\underline{f}} , \quad \hat{b}^+_{\underline{f}} = \hat{S}^-_{\underline{f}} = \hat{S}^x_{\underline{f}} - i\hat{S}^y_{\underline{f}} , \tag{3.4}$$

or

$$\hat{b}_{\underline{f}} = \begin{pmatrix} 0 & 1 \\ 0 & 0 \end{pmatrix}, \quad \hat{b}^+_{\underline{f}} = \begin{pmatrix} 0 & 0 \\ 1 & 0 \end{pmatrix} . \tag{3.5}$$

As $\hat{S}^2_{\underline{f}} = \frac{3}{4}$, only two of the components of $\hat{S}_{\underline{f}}$ are independent. From equations (3.4) and (3.5) we find

$$\hat{S}^x_{\underline{f}} = \frac{1}{2}(\hat{b}_{\underline{f}} + \hat{b}^+_{\underline{f}}) , \quad \hat{S}^y_{\underline{f}} = \frac{i}{2}(\hat{b}^+_{\underline{f}} - \hat{b}_{\underline{f}}) , \quad \hat{S}^z_{\underline{f}} = \frac{1}{2}(1 - 2\hat{n}_{\underline{f}}) , \tag{3.6}$$

where

$$\hat{n}_f = \hat{b}^+_f \hat{b}_{\underline{f}} = \begin{pmatrix} 0 & 0 \\ 0 & 1 \end{pmatrix} . \tag{3.7}$$

By substitution of equation (3.6) into equation (3.1) we get (N is the number of spins in the lattice)

$$\begin{aligned} H_{op} = & - \mu_B N B + 2\mu_B B \sum_{\underline{f}} \hat{n}_{\underline{f}} - 2 \sum_{\underline{f},\underline{g}} I(\underline{f}-\underline{g})\ \hat{b}^+_{\underline{f}}\hat{b}_{\underline{g}} - \frac{1}{2}N \sum_{\underline{f}} I(\underline{f}) \\ & + 2\Big(\sum_{\underline{f}} I(\underline{f})\Big)\sum_{\underline{g}} \hat{n}_{\underline{g}} - 2 \sum_{\underline{f},\underline{g}} I(\underline{f}-\underline{g})\ \hat{n}_{\underline{f}}\hat{n}_{\underline{g}} . \end{aligned} \tag{3.8}$$

We shall study the Green function $\langle\langle \hat{b}_{\underline{g}} ; \hat{b}^+_{\underline{f}} \rangle\rangle$ so that we first of all need the commutation relations for the $\hat{b}_{\underline{f}}$ and $\hat{b}^+_{\underline{f}}$. These are

$$\left[\hat{b}_{\underline{f}},\hat{b}_{\underline{g}}\right]_- = \left[\hat{b}^+_{\underline{f}},\hat{b}^+_{\underline{g}}\right]_- = \hat{b}^2_{\underline{f}} = \hat{b}^{+2}_{\underline{f}} = 0 , \quad \left[\hat{b}_{\underline{f}},\hat{b}^+_{\underline{g}}\right]_- = \left[1 - 2\hat{n}_{\underline{f}}\right] \delta_{\underline{fg}} . \tag{3.9}$$

From equation (1.17) with $\eta = +1$ we now get

$$E \langle\langle \hat{b}_{\underline{g}};\hat{b}^+_{\underline{f}}\rangle\rangle = \frac{1}{2\pi}\, \delta_{\underline{fg}}(1 - 2\langle\hat{n}_{\underline{f}}\rangle) + 2\left[\mu_B B + \sum_{\underline{f}} I(\underline{f})\right]\langle\langle \hat{b}_{\underline{g}};\hat{b}^+_{\underline{f}}\rangle\rangle$$

$$- 2\sum_{\underline{p}} I(\underline{p}-\underline{g})\, \langle\langle \hat{b}_{\underline{p}};\hat{b}^+_{\underline{f}}\rangle\rangle + 4\sum_{\underline{p}} I(\underline{p}-\underline{g}) \left\{\langle\langle \hat{n}_{\underline{g}}\hat{b}_{\underline{p}};\hat{b}^+_{\underline{f}}\rangle\rangle - \langle\langle \hat{n}_{\underline{p}}\hat{b}_{\underline{g}};\hat{b}^+_{\underline{f}}\rangle\rangle\right\} . \tag{3.10}$$

We note here the appearance of Green functions involving four rather than two $\hat{b}_{\underline{f}}$ or $\hat{b}^+_{\underline{f}}$. To find equations for those, we should apply equation (1.17) again, but the resultant equations involve Green functions with six $\hat{b}_{\underline{f}}$ or $\hat{b}^+_{\underline{f}}$, and so on. Sooner or later we must, therefore, in some way or other close the chain of equations. As, because of equation (3.3), the $\hat{n}_{\underline{f}}$ and $\hat{b}_{\underline{f}}$ operators in the more complex Green functions refer to different sites, we shall assume them to be uncorrelated in first approximation; that is, we put

$$\langle\langle \hat{n}_{\underline{f}}\hat{b}_{\underline{g}};\hat{b}^+_{\underline{h}}\rangle\rangle \approx \langle n_{\underline{f}}\rangle \langle\langle \hat{b}_{\underline{g}};\hat{b}^+_{\underline{h}}\rangle\rangle . \tag{3.11}$$

As our problem has translational symmetry, $\langle n_{\underline{f}}\rangle$ is independent of $\underline{f}$, and we shall write

$$\langle n_{\underline{f}}\rangle = \bar{n} . \tag{3.12}$$

From equations (3.10) and (3.11) we now get

$$E \langle\langle \hat{b}_{\underline{g}};\hat{b}^+_{\underline{f}}\rangle\rangle = \frac{1 - 2\bar{n}}{2\pi}\, \delta_{\underline{fg}} + \left[2\mu_B B + 2(1 - 2\bar{n}) \sum_{\underline{f}} I(\underline{f})\right] \langle\langle \hat{b}_{\underline{g}};\hat{b}^+_{\underline{f}}\rangle\rangle$$

$$- 2(1 - 2\bar{n}) \sum_{\underline{p}} I(\underline{p} - \underline{g})\, \langle\langle \hat{b}_{\underline{p}};\hat{b}^+_{\underline{f}}\rangle\rangle . \tag{3.13}$$

To solve this equation, we use the translational invariance once again and Fourier transform with respect to the lattice sites. We write

$$\langle\langle \hat{b}_{\underline{g}};\hat{b}^+_{\underline{f}}\rangle\rangle = \frac{1}{N} \sum_{\underline{q}\,\underline{g}} e^{i(\underline{g} - \underline{f}.\underline{q})} G_{\underline{q}} \tag{3.14}$$

and use the fact that

$$\delta_{\underline{fg}} \equiv \frac{1}{N} \sum_{\underline{q}} e^{i(\underline{g} - \underline{f}.\underline{q})} . \tag{3.15}$$

where the sums over $\underline{q}$ are over the first Brillouin zone. We now get from equation (3.13)

$$G_{\underline{q}}\left\{E-2\mu_B B-2(1-2\bar{n})\left[K(0)-K(\underline{q})\right]\right\}=\frac{1-2\bar{n}}{2\bar{n}}\ , \tag{3.16}$$

where

$$K(\underline{q})=\sum_{\underline{f}} I(\underline{f})\ e^{i(\underline{f}.\underline{q})}\ , \tag{3.17}$$

or

$$G_{\underline{q}}=\frac{1-2\bar{n}}{2\pi(E-E_{\underline{q}})}\ , \tag{3.18}$$

with

$$E_{\underline{q}}=2\mu_B B+2(1-2\bar{n})\left[K(0)-K(\underline{q})\right]. \tag{3.19}$$

From equations (3.18), (3.14), (1.30), (1.28) and (1.19) we now get

$$\langle b^+_{\underline{f}}(t')b_{\underline{g}}(t)\rangle=\frac{1-2\bar{n}}{N}\sum_{\underline{q}}\frac{\exp[i(\underline{g}-\underline{f}.\underline{q})-iE_{\underline{q}}(t-t')]}{e^{\beta E_{\underline{q}}}-1}. \tag{3.20}$$

Putting $t=t'$ and $\underline{f}=\underline{g}$, using equations (3.7) and (3.12), and changing from a summation over $\underline{q}$ to an integration,

$$\frac{1}{N}\sum_{\underline{q}}\ldots\rightarrow\frac{v}{(2\pi)^3}\int d^3q\ \ldots\ ,\quad v=\frac{V}{N}\ , \tag{3.21}$$

we get the following implicit equation for $\bar{n}$:

$$\frac{\bar{n}}{1-2\bar{n}}=\frac{v}{(2\pi)^3}\int\frac{d^3\underline{q}}{e^{\beta E_{\underline{q}}}-1}. \tag{3.22}$$

The physical quantity usually measured is the magnetization which is proportional to $\langle S^z\rangle$ and thus to $1-2\bar{n}=\sigma$. Using the fact that the number of $\underline{q}$-values within the first Brillouin zone is N, we have

$$1=\frac{v}{(2\pi)^3}\int d^3\underline{q}\ , \tag{3.23}$$

and hence we get from equation (3.22)

$$\frac{1}{\sigma}=\frac{v}{(2\pi)^3}\int\coth\left(\tfrac{1}{2}\beta E_{\underline{q}}\right)d^3\underline{q}\ . \tag{3.24}$$

It is interesting to see how good an approximation we have obtained. Let us first see where the Curie point T_c occurs. Let this be for $\beta=\beta_c$. Near, but below, T_c the quantity σ, and thus E_q, will be small (we must put $B=0$) We can thus expand the coth in a power series; hence

$$\frac{1}{\sigma}=\frac{v}{(2\pi)^3}\int\left\{\frac{2}{\sigma\beta K(0)\eta(\underline{q})}+\frac{\sigma\beta K(0)\eta(\underline{q})}{6}-\frac{\sigma^3\beta^3K^3(0)\eta^3(\underline{q})}{180}+\ldots\right\}d^3\underline{q}\ , \tag{3.25}$$

where

$$\eta(\underline{q}) = 1 - \frac{K(\underline{q})}{K(0)} \quad . \tag{3.26}$$

For simple (sc), body-centred (bcc), and face-centred (fcc) cubic lattices we have for the case of nearest neighbour interactions

$$\left.\begin{aligned} K(\underline{q}) &= \tfrac{1}{3}K(0)\left[\cos q_x a + \cos q_y a + \cos q_z a\right] , && \text{(sc)} \\ K(\underline{q}) &= K(0) \cos q_x a \, \cos q_y a \, \cos q_z a \quad , && \text{(bcc)} \\ K(\underline{q}) &= \tfrac{1}{3}K(0)\left[\cos(\tfrac{1}{2}q_x a)\cos(\tfrac{1}{2}q_y a) + \cos(\tfrac{1}{2}q_y a)\cos(\tfrac{1}{2}q_z a)\right. \\ &\qquad \left. + \cos(\tfrac{1}{2}q_z a)\cos(\tfrac{1}{2}q_x a)\right] , && \text{(fcc)} \end{aligned}\right\} \tag{3.27}$$

where $a\sqrt{6/z}$ is the nearest neighbour distance, and z the coordination number ($z = 6$, sc; $z = 8$, bcc; $z = 12$, fcc).

If we define

$$F(n) = \frac{v}{(2\pi)^3} \int \eta^n(\underline{q}) \, d^3\underline{q} \ , \tag{3.28}$$

the $F(n)$ depend on the crystal structure and have been computed. We have

$$F(-1) = 1.56 \text{ (sc)} \ , = 1.393 \text{ (bcc)} \ , = 1.345 \text{ (fcc)} \ ; \tag{3.29}$$

$$F(1) = 1 \ , \ F(2) = \frac{z+1}{z} \ , \ F(3) = \frac{z+3}{z} \ , \ \ldots \quad . \tag{3.30}$$

From equation (3.25) we now get

$$\frac{1}{\sigma} = \frac{2F(-1)}{\sigma\beta K(0)} + \frac{1}{6}\sigma\beta K(0) - \ldots \quad . \tag{3.31}$$

By letting σ tend to zero, we find

$$\beta_c = \frac{2F(-1)}{K(0)} = F(-1)\,\beta_{c_0} \quad , \tag{3.32}$$

where β_{c_0} is the molecular field result; β_c is, in fact the same as the spherical model result. For $\beta - \beta_c \ll \beta_c$ we get a temperature dependence of σ which is the same as in the molecular field theory (and different from the experimental T-dependence).

$$\sigma = \sqrt{\frac{6}{F(-1)}}\sqrt{1 - \frac{T}{T_c}} \quad . \tag{3.33}$$

At low temperatures (and still $B = 0$), it is more convenient to work with equation (3.22), as now $\sigma \approx 1$. We note that the small q-values will dominate so that we can integrate over the whole of q-space. If we write $\bar{n}/(1 - 2\bar{n}) = \Phi$, we get

$$\Phi = \frac{v}{(2\pi)^3}\int_0^{2\pi} d\varphi \int_0^{\pi} \sin\theta \, d\theta \int_0^{q} q^2 dq \sum_{r=1}^{\infty} \exp\left[-2r\,\beta\sigma K(0)\,\eta(\underline{q})\right] . \tag{3.34}$$

Expanding $\eta(\underline{q})$ in powers of $\underline{q}$ and integrating, we get a power series in β^{-1} or T and the result is

$$\Phi = \zeta(\tfrac{3}{2})(\tfrac{1}{2}\kappa)^{\frac{3}{2}} + \frac{3\pi}{4}\nu\zeta(\tfrac{5}{2})(\tfrac{1}{2}\kappa)^{\frac{5}{2}} + \pi^2\omega\nu^2\zeta(\tfrac{7}{2})(\tfrac{1}{2}\kappa)^{\frac{7}{2}} + \ldots \tag{3.35}$$

where $\zeta(n)$ is the Riemann zeta function ,

$$\kappa = \frac{3}{2\pi\nu\,\beta K(0)}, \tag{3.36}$$

and

$$\nu = 1,\ \omega = \frac{33}{32},\ \mathrm{sc}\ ;\quad \nu = 2^{-\frac{4}{3}}.3,\ \omega = \frac{281}{288},\ \mathrm{bcc}\ ;\ \nu = 2^{\frac{1}{3}},\ \omega = \frac{15}{16},\ \mathrm{fcc}\ . \tag{3.37}$$

From equation (3.35) we find for σ a power series in T :

$$\sigma = 1 - a_0 T^{\frac{3}{2}} - a_1 T^{\frac{7}{2}} \ldots - b_0 T^3 - b_1 T^4 - \ldots\quad . \tag{3.38}$$

This result agrees with the exact series expansion up to the terms in $T^{\frac{5}{2}}$, but in the exact series there is no term in T^3. Part of the trouble lies in the fact that the temperature dependence of the $E_{\underline{q}}$, the elementary excitation (that is, spin-wave) energies, is incorrect. From equation (3.18) and equation (3.28) one finds that

$$E_{\underline{q}}(T) = E_{\underline{q}}(0) + B\,T^{\frac{3}{2}}, \tag{3.39}$$

while the correct expression is

$$E_{\underline{q}}(T) = E_{\underline{q}}(0) + A\,T^{\frac{5}{2}}\ . \tag{3.40}$$

Let us finally see what equation (3.24) leads to at high temperatures where now $B \neq 0$, as otherwise $\sigma = 0$. We write

$$\coth(\tfrac{1}{2}\beta E_{\underline{q}}) = \frac{1 + t_0 t_1}{t_0 + t_1}, \tag{3.41}$$

where

$$t_0 = \tanh(\beta\mu_B B)\ ;\quad t_1 = \tanh\left[\beta\sigma K(0)\eta(\underline{q})\right], \tag{3.42}$$

and expand equation (3.41) in powers of t_1 :

$$\coth(\tfrac{1}{2}\beta E_{\underline{q}}) = \frac{1}{t_0}\left[1 + (1 - t_0^2)\sum_n\left(\frac{-t_1}{t_0}\right)^n\right], \tag{3.43}$$

We then expand t_1 in powers of β, and using equation (3.28) and equation (3.30) we find

$$\Phi = \frac{1}{2t_0}\left[1 + \frac{\beta K(0)}{t_0} + \frac{z+1}{z}\ \frac{\beta^2 K^2(0)}{t_0^2} + \ldots\right]. \tag{3.44}$$

From this we get for the susceptibility χ per spin

$$\chi = \beta\mu_B^2 \left[1 + \frac{\beta}{\beta_{c_0}} + \frac{z-1}{z}\left(\frac{\beta}{\beta_{c_0}}\right)^2 + \dots\right], \tag{3.45}$$

which can approximately be written in the Curie-Weiss (molecular field) form

$$\chi = \frac{\mu_B^2}{k_B(T - T_{c_0})} . \tag{3.46}$$

We also note that equation (3.45) is nearly the same as the exact series in β, which was determined by Opechowski using standard thermodynamic perturbation theory.

2.4 Ferromagnetic Resonance

When studying ferromagnetic resonance we must take into account the finite shape of the sample on which resonance experiments are performed. This can be done most simply, even though only approximately, by adding to the Hamiltonian a term involving the demagnetization factors N_x, N_y and N_z, which depend on the shape of the sample, which we assume to be ellipsoidal with the principal axes along the x- , y- , and z-axes. The Hamiltonian equation (3.1) must now be supplemented by a term

$$\tfrac{1}{2}(N_x\hat{M}_x^2 + N_y\hat{M}_y^2 + N_z\hat{M}_z^2) , \tag{4.1}$$

where $\underline{M}_{op}$ is the magnetization vector ,

$$\underline{M}_{op} = 2\mu_B \sum_{\underline{f}} \hat{\underline{S}}_{\underline{f}} \tag{4.2}$$

In terms of $\hat{b}_{\underline{f}}$ and $\hat{b}_{\underline{f}}^+$ the extra term equation (4.1) has the form

$$\tfrac{1}{2}N^2N_z\mu_B^2 - 2NN_z\mu_B^2\sum_{\underline{f}}\hat{n}_{\underline{f}} + (N_x + N_y)\mu_B^2\sum_{\underline{f},\underline{g}}\hat{b}_{\underline{f}}^+\hat{b}_{\underline{g}}$$
$$+ N_z\mu_B^2\sum_{\underline{f},\underline{g}}\hat{n}_{\underline{f}}\hat{n}_{\underline{g}} + \left[\tfrac{1}{2}\mu_B^2(N_x - N_y)\sum_{\underline{f},\underline{g}}(\hat{b}_{\underline{f}}\hat{b}_{\underline{g}} + \hat{b}_{\underline{f}}^+\hat{b}_{\underline{g}}^+)\right] . \tag{4.3}$$

The term in the square brackets is the one which will introduce some new element in our calculations.

Let us now consider the experimental situation. Apart from the constant field B along the z-axis there will be a radio-frequency field $\underline{B}_1$ with components $b\cos\omega t$, $b\sin\omega t$, 0 in the xy-plane. The perturbing Hamiltonian H'_{op} will be of the form

$$H'_{op} = -2\mu_B\sum_{\underline{f}}(\hat{\underline{S}}_{\underline{f}}\cdot\underline{B}_1) = U_{op}\,e^{i\omega t} + U_{op}^*\,e^{-i\omega t} , \tag{4.4}$$

where

$$U_{op} = -b\mu_B \sum_{\underline{f}} \hat{S}_{\underline{f}}^- . \tag{4.5}$$

If $\delta\underline{M}_{op}$ is the additional magnetization in the x y-plane produced by B_1 we write

$$\delta\hat{M}_\pm = \delta(\hat{M}_x \pm i\hat{M}_y) = N\chi_\pm b\, e^{\pm i\omega t} . \tag{4.6}$$

Using equation (2.11) we find for $\chi_\pm$

$$\chi_-^* = \chi_+ = -4\pi\mu_B^2 \sum_{\underline{f}} \langle\langle \hat{S}_{\underline{g}}^+ ; \hat{S}_{\underline{f}}^- \rangle\rangle . \tag{4.7}$$

If we write

$$\chi_\pm = \chi' \pm i\chi'' , \tag{4.8}$$

and use the fact that the energy absorbed per unit time, W, is given by the equation

$$W = (\underline{B}_1 \cdot \delta\dot{\underline{M}}) = \tfrac{1}{2} i\, b^2\omega(\chi_+ - \chi_-) = Nb^2\omega\chi'' , \tag{4.9}$$

we see that χ'' determines the absorption.

From equation (4.7) we see that once again we must determine $\langle\langle \hat{b}_{\underline{g}} ; \hat{b}_{\underline{f}}^+ \rangle\rangle$. However, if equation (4.3) is added to the Hamiltonian we now find for $\langle\langle \hat{b}_{\underline{g}} ; \hat{b}_{\underline{f}}^+ \rangle\rangle$ an equation involving also $\langle\langle \hat{b}_{\underline{g}}^+ ; \hat{b}_{\underline{f}}^+ \rangle\rangle$. If we also write down the equation for this Green function, we get two coupled equations which we can solve provided we use the decoupling equation (3.11) as well as the analogous equation

$$\langle\langle \hat{n}_{\underline{f}} \hat{b}_{\underline{g}}^+ ; \hat{b}_{\underline{h}}^+ \rangle\rangle = \langle n_{\underline{f}} \rangle \langle\langle \hat{b}_{\underline{g}}^+ ; \hat{b}_{\underline{h}}^+ \rangle\rangle . \tag{4.10}$$

As a result we obtain the following equations

$$\begin{aligned} E\langle\langle \hat{b}_{\underline{g}} ; \hat{b}_{\underline{f}}^+ \rangle\rangle = {} & \frac{\sigma}{2\pi}\delta_{\underline{fg}} + \left[2\mu_B B - 2NN_z\mu_B^2 + 2\sigma K(0)\right]\langle\langle \hat{b}_{\underline{g}} ; \hat{b}_{\underline{f}}^+ \rangle\rangle \\ & + \sum_{\underline{p}}\left[-2\sigma I(\underline{g}-\underline{p}) + \sigma\mu_B^2(N_x+N_y)\right]\langle\langle \hat{b}_{\underline{p}} ; \hat{b}_{\underline{f}}^+ \rangle\rangle \\ & + \sigma\mu_B^2(N_x - N_y)\sum_{\underline{p}}\langle\langle \hat{b}_{\underline{p}}^+ ; \hat{b}_{\underline{f}}^+ \rangle\rangle , \end{aligned} \tag{4.11}$$

$$\begin{aligned} E\langle\langle \hat{b}_{\underline{g}}^+ ; \hat{b}_{\underline{f}}^+ \rangle\rangle = {} & -\left[2\mu_B^2 B - 2NN_z\mu_B^2 + 2\sigma K(0)\right]\langle\langle \hat{b}_{\underline{g}}^+ ; \hat{b}_{\underline{f}}^+ \rangle\rangle \\ & - \sum_{\underline{p}}\left[-2\sigma I(\underline{g}-\underline{p}) + \sigma\mu_B^2(N_x+N_y)\right]\langle\langle \hat{b}_{\underline{p}}^+ ; \hat{b}_{\underline{f}}^+ \rangle\rangle - \sigma\mu_B^2(N_x-N_y)\langle\langle \hat{b}_{\underline{p}} ; \hat{b}_{\underline{p}}^+ \rangle\rangle . \end{aligned} \tag{4.12}$$

Once again we Fourier transform with respect to the lattice sites and use equation (3.14) as well as

$$\langle\langle \hat{b}^+_{\underline{g}};\hat{b}^+_{\underline{f}}\rangle\rangle = \frac{1}{N}\sum_{\underline{q}} \Gamma_{\underline{q}}\, e^{i(\underline{g}-\underline{f}.\underline{q})} \tag{4.13}$$

and

$$\frac{1}{N}\sum_{\underline{f}} e^{i(\underline{f}.\underline{q})} = \delta_{\underline{q}o} \quad . \tag{4.14}$$

The result is

$$(E-\bar{E}_{\underline{q}})G_{\underline{q}} = \frac{\sigma}{2\pi} + \sigma\mu_B^2(N_x-N_y)\,N\,\delta_{\underline{q}o}\,\Gamma_{\underline{q}} \quad , \tag{4.15}$$

$$(E+\bar{E}_{\underline{q}})\Gamma_{\underline{q}} = -\sigma\mu_B^2(N_x-N_y)\,N\,\delta_{\underline{q}o}\,G_{\underline{q}} \quad , \tag{4.16}$$

with

$$\bar{E}_{\underline{q}} = E_{\underline{q}} + N\sigma\mu_B^2\left[(N_x+N_y)\,\delta_{\underline{q}o} - 2N_z\right] \quad . \tag{4.17}$$

We note that if $\underline{q} \neq 0$, $\bar{E}_{\underline{q}}$ differs from $E_{\underline{q}}$ in that B is replaced by $B - N_z\,(N\sigma\mu_B)$, that is, by the field including the demagnetization.

Solving equation (4.15) and equation (4.16) we get

$$\underline{q} \neq 0 : \Gamma_{\underline{q}} = 0 \;, \quad G_{\underline{q}} = \frac{\sigma}{2\pi(E-\bar{E}_{\underline{q}})} \tag{4.18}$$

$$\underline{q} = 0 : G_o = \frac{\sigma}{2\pi}\,\frac{E+E_o}{E^2-E_r^2} \quad , \tag{4.19}$$

where

$$E_r^2 = 4\mu_B^2\left[B+N\sigma\mu_B(N_x-N_z)\right]\left[B+N\sigma\mu_B(N_y-N_z)\right] . \tag{4.20}$$

From equations (4.7), (3.4), (3.14), (4.18) and (4.19) we get for χ_+ :

$$\chi_+ = -4\pi\mu_B^2\sum_{\underline{f}}\langle\langle \hat{b}_{\underline{g}};\hat{b}^+_{\underline{f}}\rangle\rangle = -\frac{4\pi\mu_B^2}{N}\sum_{\underline{f}}\sum_{\underline{g}} G_{\underline{q}}\,e^{i(\underline{f}-\underline{g}.\underline{q})}$$

$$= \lim_{\epsilon\to 0+}\left[-\frac{4\pi\mu_B^2}{N}\sum_{\underline{q}\neq 0}\frac{\sigma}{2\pi}\,\frac{1}{\omega+i\epsilon-E_{\underline{q}}}\sum_{\underline{f}} e^{i(\underline{g}-\underline{f}.\underline{q})} - \frac{4\pi\mu_B^2}{N}\sum_{\underline{f}}\frac{\sigma}{2\pi}\,\frac{\omega+i\epsilon-\bar{E}_o}{(\omega+i\epsilon)^2-E_r^2}\right.$$

$$= \lim_{\epsilon\to 0+}\left[-\frac{\sigma\mu_B^2}{E_r}(\omega+i\epsilon-\bar{E}_o)\left\{\frac{1}{\omega+i\epsilon-E_r} - \frac{1}{\omega+i\epsilon+E_r}\right\}\right] \quad , \tag{4.21}$$

and using equation (1.28), and the fact that both ω and E_r are positive so that $\delta(\omega+E_r)=0$, we get for χ''

$$\chi'' = \frac{2\pi\mu_B^2(\omega+\bar{E}_o)}{E_r}\,\delta(\omega-E_r) \quad , \tag{4.22}$$

which shows that absorption takes place only at $\omega = E_r$. In our present approximation we find zero line-width. If we take into account s-d or spin-phonon coupling or go to a higher-order decoupling, we obtain both a finite linewidth and a line-shift.

2.5 Antiferromagnetics

We now consider the following Hamiltonian

$$H_{op} = \sum_{\underline{f},\underline{g}} I(\hat{S}_{\underline{f}}.\hat{S}_{\underline{g}}) - 2\mu_B B\left[\sum_{\underline{f}} \hat{S}^z_{\underline{f}} + \sum_{\underline{g}} \hat{S}^z_{\underline{g}}\right] , \tag{5.1}$$

where we have assumed the existence of two sublattices such that all nearest neighbours of a spin on one sublattice (indicated by $\underline{f}$) are on the other sublattice (indicated by $\underline{g}$) and where the double sum is over all nearest neighbour pairs only. In other words, we have put in equation (3.1) $I(\underline{f}-\underline{g}) = 0$, unless $\underline{f}$ and $\underline{g}$ are nearest neighbours, in which case $I(\underline{f}-\underline{g}) = -I$. In terms of the $\hat{b}_{\underline{f}}$ and $\hat{b}^+_{\underline{f}}$, equation (5.1) becomes, similar to equation (3.8) :

$$\hat{H} = -\mu_B B N + 2\mu_B B\left[\sum_{\underline{f}} \hat{n}_{\underline{f}} + \sum_{\underline{g}} \hat{n}_{\underline{g}}\right] - 2\sum_{\underline{f},\underline{g}} I\, \hat{b}^+_{\underline{f}}\hat{b}_{\underline{g}} - \tfrac{1}{2} z N I$$

$$+ 2zI\left[\sum_{\underline{f}} \hat{n}_{\underline{f}} + \sum_{\underline{g}} \hat{n}_{\underline{g}}\right] - 2I \sum_{\underline{f},\underline{g}}{}' \hat{n}_{\underline{f}}\, \hat{n}_{\underline{g}} . \tag{5.2}$$

We now consider the Green functions $\langle\langle \hat{b}_{\underline{f}};\hat{b}^+_{\underline{h}}\rangle\rangle$ and $\langle\langle \hat{b}_{\underline{g}};\hat{b}^+_{\underline{h}}\rangle\rangle$, where $\underline{h}$ may refer to either the $\underline{f}$- or the $\underline{g}$- sublattice. The equations of motion are ($\underline{\ell}$ is either $\underline{f}$ or $\underline{g}$; compare equation (3.10))

$$E\langle\langle \hat{b}_{\underline{\ell}};\hat{b}^+_{\underline{h}}\rangle\rangle = \frac{\delta_{\underline{\ell}\underline{h}}}{2\pi}\langle 1 - 2\hat{n}_{\underline{\ell}}\rangle + 2(\mu_B B + zI)\langle\langle \hat{b}_{\underline{\ell}};\hat{b}^+_{\underline{h}}\rangle\rangle$$
$$- 2I\sum_{\underline{k}}\langle\langle \hat{b}_{\underline{k}};\hat{b}^+_{\underline{h}}\rangle\rangle + 4I\sum_{\underline{k}}\langle\langle \hat{n}_{\underline{\ell}}\hat{b}_{\underline{k}} - \hat{n}_{\underline{k}}\hat{b}_{\underline{\ell}} ; \hat{b}^+_{\underline{h}}\rangle\rangle$$

where the sums over $\underline{k}$ are over the z nearest neighbours of $\underline{\ell}$. We use again the decoupling equation (3.11), but bear in mind that now $\langle n_{\underline{f}}\rangle$ is not necessarily the same as $\langle n_{\underline{g}}\rangle$. Proceeding just as in subsection 2.3, we Fourier transform with respect to the lattice. To fix our ideas, let $\underline{h}$ belong to the $\underline{f}$-sublattice so that $\delta_{\underline{g}\underline{h}} \equiv 0$, and let G denote $\langle\langle \hat{b}_{\underline{f}} ; \hat{b}^+_{\underline{h}}\rangle\rangle$ and Γ denote $\langle\langle \hat{b}_{\underline{g}};\hat{b}_{\underline{h}}\rangle\rangle$. After some calculations, we find for G and Γ the following equations

$$G = \frac{\sigma_{\underline{f}}}{4\pi N}\sum_{\underline{q}} e^{i(\underline{q}.\underline{f}-\underline{h})}\,\frac{E - 2\mu_B B + 2zI\sigma_{\underline{f}}}{(E-E_1)\ (E-E_2)} , \tag{5.4}$$

$$\Gamma = \frac{I\sigma_{\underline{g}}\,\sigma_{\underline{f}} z}{2\pi N}\sum_{\underline{q}} e^{i(\underline{q}.\underline{f}-\underline{h})}\,\frac{\gamma_{\underline{q}}}{(E-E_1)(E-E_2)} , \tag{5.5}$$

where

$$\gamma_{\underline{q}} = \frac{1}{z}\sum_{\delta} e^{i(\underline{\delta}.\underline{q})} , \tag{5.6}$$

$\underline{\delta} = \underline{f}-\underline{g}$ for $\underline{f}$ and $\underline{g}$ being nearest neighbours, and the summation is over all nearest neighbours,

$$\sigma_{\underline{f},\underline{g}} = 1 - 2\langle n_{\underline{f},\underline{g}}\rangle , \tag{5.7}$$

and E_1 and E_2 are the solutions of the equation

$$\left[E - 2\mu_B B + 2I z\sigma_{\underline{f}}\right]\left[E - 2\mu_B B + 2I z\sigma_{\underline{g}}\right] - 4I^2 \sigma_{\underline{f}}\sigma_{\underline{g}} z^2\gamma_{\underline{q}}^2 = 0 . \tag{5.8}$$

The sublattice magnetizations $\sigma_{\underline{f},\underline{g}}$ follow from G and $\langle\langle \hat{b}_{\underline{g}};\hat{b}^+_{\underline{g}'}\rangle\rangle$, and we find

$$\sigma_{\underline{f},\underline{g}} = \left[1 + 2\Psi_{\underline{f},\underline{g}}\right]^{-1} , \tag{5.9}$$

where

$$\Psi_{\underline{f},\underline{g}} = \frac{1}{2N}\sum_{\underline{q}}\left[f(1) + f(2) \pm \frac{2I z(\sigma_{\underline{f}}-\sigma_{\underline{g}})}{E_1 - E_2}\left\{f(1) - f(2)\right\}\right] , \tag{5.10}$$

where

$$f(i) = \left[e^{\beta E_i} - 1\right]^{-1} . \tag{5.11}$$

Let $\sigma_{\underline{f}} = -\sigma_{\underline{g}} = \sigma_0$ in the absence of a magnetic field. We then have

$$\Psi_{\underline{f}} = \frac{1}{2}\left[\frac{1}{N}\sum_{\underline{q}} \frac{2I z\sigma_0}{E_0}\coth(\tfrac{1}{2}\beta E_0) - 1\right] , \tag{5.12}$$

with

$$E_0 = 2\sigma_0 I(1 - \gamma_{\underline{q}})^{\frac{1}{2}} . \tag{5.13}$$

As $T \to T_N$ (T_N is the Néel temperature), $\Psi_{\underline{f}} \to \infty$ and

$$\sigma_0 \sim \frac{1}{4\Psi_{\underline{f}} + 2} + \dots , \tag{5.14}$$

and

$$\Psi_{\underline{f}} \approx \frac{1}{N}\sum_{\underline{q}} \frac{k_B T}{2I z\sigma_0(1 - \gamma_{\underline{q}}^2)} + \dots . \tag{5.15}$$

By considering the limit as $\sigma_0 \to 0$ we find T_N:

$$T_N = \frac{I z}{2k_B}\left[\frac{1}{N}\sum_{\underline{q}} \frac{1}{1 - \gamma_q}\right]^{-1} , \tag{5.16}$$

which is the same as T_c given by equation (3.32).

The zero-field parallel susceptibility follows from the relation

$$\chi_{\parallel}^{(0)} = \lim_{B\to 0} \frac{\sigma_{\underline{f},\underline{g}} \mp \sigma_0}{B} \quad , \tag{5.17}$$

whence we find

$$\chi_{\parallel}^{(0)} = \lim_{B\to 0} \sum_{i=\underline{f},\underline{g}} \left\{ \frac{\Psi_i(1+\Psi_i)}{(1+2\Psi_i)^2} - \frac{1}{2} \right\} \frac{\partial \Psi_i}{\partial B} \quad . \tag{5.18}$$

For $T \geq T_N$ one finds

$$\chi_{\parallel}^{(0)} = \frac{3}{4k_B T} \left[\frac{1}{N} \sum_{\underline{q}} \frac{1}{4\mu_B^2 - 2 I z \chi_{\parallel}^0 (1+\gamma_{\underline{q}})} \right]^{-1} \quad , \tag{5.19}$$

which at $T = T_N$ leads to the molecular field value

$$\chi(T_N) = \frac{\mu_B^2}{Iz} \quad . \tag{5.20}$$

One can also evaluate $\chi_{\parallel}$ below T_N and one can evaluate the perpendicular susceptibility, for which one applies the fluctuation-dissipation theorem.

2.6 The Paramagnetic Phase of a Heisenberg Ferromagnet

In the present subsection we shall be concerned with the Heisenberg ferromagnet in the temperature region above the critical point — the paramagnetic region — in zero external magnetic field. This region has come more into the focus of attention because of recent interest in phase transitions and studies of physical behaviour at both sides of transition temperatures. The results obtained so far for this region can be summarized as follows: From series expansions in the inverse absolute temperature T^{-1} it is found that the susceptibility is positive, vanishes at T^{-1} at infinite temperature, and diverges as $(T - T_c)^{-\frac{4}{3}}$ at the Curie temperature T_c. Approximate methods such as the Bogolyubov-Tyablikov decoupling scheme used in subsection 2.3 (called in the following the RPA (random phase approximation) theory) give the correct asymptotic behaviour as $T \to \infty$, but a $(T - T_c)^{-2}$ behaviour for $T \to T_c$, a behaviour also found in the spherical model.

Brout has suggested that excitations corresponding to short-range order correlations might exist in the paramagnetic region. These excitations would be 'quasi-spin-waves' with wavelengths short compared with the correlation length. A theoretical treatment by Beeby and Hubbard found evidence for such quasi-excitations, and Bennett has recently predicted, from considering the generalized susceptibility, propagating excitations at wave-numbers k greater than a lower limit k_c, which is proportional to the square root of the inverse susceptibility. The experimental evidence for such quasi-excitations is by now considerable, as well.

For the Hamiltonian of the system we choose

$$H_{op} = -\frac{1}{2} \sum_{\underline{a},\underline{b}} J_{\underline{a}\,\underline{b}} (\hat{S}_{\underline{a}} \cdot \hat{S}_{\underline{b}}) , \quad J_{\underline{a}\,\underline{a}} = 0 , \tag{6.1}$$

where $\underline{a}$ and $\underline{b}$ label the sites of a lattice of N spins of magnitude S and $J_{\underline{a}\,\underline{b}} (= J_{\underline{b}\,\underline{a}})$ is an exchange energy. In subsection 2.3 we used the decoupling equation (3.11) which essentially meant

$$\langle\langle \hat{S}^z_{\underline{c}} \hat{S}^+_{\underline{a}} ; \hat{S}^-_{\underline{b}} \rangle\rangle \cong \langle \hat{S}^z \rangle \langle\langle \hat{S}^+_{\underline{a}} ; \hat{S}^-_{\underline{b}} \rangle\rangle . \tag{6.2}$$

The quantity $\langle \hat{S}^z \rangle$ measures the long-range order (magnetization), and as long as it is non-vanishing it whould be the main determing factor for the behaviour of the system. However, in the paramagnetic phase where there is no external field, $\langle \hat{S}^z \rangle = 0$ and the decoupling equation (6.2) ceases to be useful. We must, therefore, look for a different kind of decoupling.

We consider the following dyadics (indicated by doubly underlined roman type) :

$$\underline{\underline{G}}^{(1)}_{\underline{f}\,\underline{g}} = \langle\langle \hat{\underline{S}}_{\underline{f}} ; \hat{\underline{S}}_{\underline{g}} \rangle\rangle . \tag{6.3}$$

We use the elementary commutation relations for spin operators

$$\left[\hat{S}^i_{\underline{a}} , \hat{S}^j_{\underline{b}} \right]_- = i\, \epsilon_{ijk}\, \hat{S}^k_{\underline{a}}\, \delta_{\underline{a}\,\underline{b}} , \tag{6.4}$$

where ϵ_{ijk} is the antisymmetric third-rank tensor :

$$\left.\begin{aligned} \epsilon_{ijk} &= 1 \quad \text{if ijk is an even permutation of 123} , \\ &= -1 \quad \text{if ijk is an odd permutation of 123} , \\ &= 0 \quad \text{otherwise} , \end{aligned}\right\} \tag{6.5}$$

where i , j and k are Cartesian coordinates.

We then find the equations of motion

$$\omega \langle\langle \hat{\underline{S}}_{\underline{f}} ; \hat{\underline{S}}_{\underline{g}} \rangle\rangle = \langle [\hat{\underline{S}}_{\underline{f}} , \hat{\underline{S}}_{\underline{g}}]_- \rangle + i \sum_{\underline{a}} J_{\underline{f}\,\underline{a}} \langle\langle [\hat{\underline{S}}_{\underline{f}} \times \hat{\underline{S}}_{\underline{a}}] ; \hat{\underline{S}}_{\underline{g}} \rangle\rangle . \tag{6.6}$$

By taking the (double) scalar product $(\underline{e}_x + i\underline{e}_y \cdot \langle\langle \hat{\underline{S}}_{\underline{f}}) ; (\hat{\underline{S}}_{\underline{g}} \rangle\rangle \cdot \underline{e}_x - i\underline{e}_y)$, where $\underline{e}_x$ and $\underline{e}_y$ are unit vectors along the x- and y-axes, we find from equation (6.6) the usual RPA equation. As we cannot use the RPA decoupling, we must go to the next order and write down the equation of motion for $\langle\langle [\hat{\underline{S}}_{\underline{f}} \times \hat{\underline{S}}_{\underline{h}}]; \hat{\underline{S}}_{\underline{g}} \rangle\rangle$, which is

$$\omega \langle\langle [\hat{\underline{S}}_{\underline{f}} \times \hat{\underline{S}}_{\underline{h}}] ; \hat{\underline{S}}_{\underline{g}} \rangle\rangle = \langle \left[[\hat{\underline{S}}_{\underline{f}} \times \hat{\underline{S}}_{\underline{h}}] , \hat{\underline{S}}_{\underline{g}} \right]_- \rangle + i \sum_{\underline{a}} \left\{ J_{\underline{f}\,\underline{a}} \langle\langle \left[[\hat{\underline{S}}_{\underline{f}} \times \hat{\underline{S}}_{\underline{a}}] \times \hat{\underline{S}}_{\underline{h}} \right] ; \hat{\underline{S}}_{\underline{g}} \rangle\rangle + J_{\underline{h}\,\underline{a}} \langle\langle \left[\hat{\underline{S}}_{\underline{f}} \times [\hat{\underline{S}}_{\underline{h}} \times \hat{\underline{S}}_{\underline{a}}] \right] ; \hat{\underline{S}}_{\underline{g}} \rangle\rangle \right\} . \tag{6.7}$$

We now expand the triple vector products by using a standard identity for classical vectors, in which additional terms appear due to the non-commutativity of the spin operators. If we order the operators so as to minimize the number of commutator terms that appear, and use the fact that $J_{\underline{a}\,\underline{a}} = 0$, we have from equation (6.7)

$$\begin{aligned} \omega \langle\langle [\hat{\underline{S}}_{\underline{f}} \times \hat{\underline{S}}_{\underline{h}}] ; \hat{\underline{S}}_{\underline{g}} \rangle\rangle = {} & \langle \left[[\hat{\underline{S}}_{\underline{f}} \times \hat{\underline{S}}_{\underline{h}}] , \hat{\underline{S}}_{\underline{g}} \right]_- \rangle \\ & + i \sum_{\underline{a}} \Big\{ J_{\underline{f}\,\underline{a}} \left\{ \langle\langle \hat{\underline{S}}_{\underline{a}} (\hat{\underline{S}}_{\underline{f}} \cdot \hat{\underline{S}}_{\underline{h}}) ; \hat{\underline{S}}_{\underline{g}} \rangle\rangle - \langle\langle \hat{\underline{S}}_{\underline{f}} (\hat{\underline{S}}_{\underline{a}} \cdot \hat{\underline{S}}_{\underline{h}}) ; \hat{\underline{S}}_{\underline{g}} \rangle\rangle \right\} \\ & - i J_{\underline{h}\,\underline{a}} \left\{ \langle\langle (\hat{\underline{S}}_{\underline{f}} \cdot \hat{\underline{S}}_{\underline{a}}) \hat{\underline{S}}_{\underline{h}} ; \hat{\underline{S}}_{\underline{g}} \rangle\rangle - \langle\langle (\hat{\underline{S}}_{\underline{f}} \cdot \hat{\underline{S}}_{\underline{h}}) \hat{\underline{S}}_{\underline{a}} ; \hat{\underline{S}}_{\underline{g}} \rangle\rangle \right\} \Big\} . \end{aligned} \tag{6.8}$$

The commutator terms do not appear at all in this equation, and thus we may say that it expresses a sort of semi-classical approximation.

We now look for a decoupling to get a closed set of equations. As we are interested in correlations in the paramagnetic region, which means that ultimately we wish to calculate (self-consistently) the correlation functions related to the $\underline{\underline{G}}^{(1)}_{\underline{f}\,\underline{g}}$, we try the decoupling

$$\langle\langle \hat{\underline{S}}_{\underline{a}} (\hat{\underline{S}}_{\underline{f}} \cdot \hat{\underline{S}}_{\underline{h}}) ; \hat{\underline{S}}_{\underline{g}} \rangle\rangle \cong \langle (\hat{\underline{S}}_{\underline{f}} \cdot \hat{\underline{S}}_{\underline{h}}) \rangle \langle\langle \hat{\underline{S}}_{\underline{a}} ; \hat{\underline{S}}_{\underline{g}} \rangle\rangle , \tag{6.9}$$

and so on. We note (i) the similarity between equations (6.2) and (6.9), (ii) that other averages such as $\langle \hat{S}^x_{\underline{a}} \hat{S}^x_{\underline{f}} \rangle$, which we might extract from the operator $\hat{\underline{S}}_{\underline{a}} (\hat{\underline{S}}_{\underline{f}} \cdot \hat{\underline{S}}_{\underline{h}})$, would cancel terms arising from $\hat{\underline{S}}_{\underline{f}} (\hat{\underline{S}}_{\underline{a}} \cdot \hat{\underline{S}}_{\underline{h}})$, while averages such as $\langle \hat{S}^x_{\underline{a}} \hat{S}^y_{\underline{f}} \rangle$ vanish in a system where the total z component of spin is conserved, and (iii) that as usual we do not give any proof of the validity of the decoupling, but hope that the results will justify our procedure.

We shall use equation (6.9) and the short-hand notation equation (6.3) and

$$\langle\langle [\hat{\underline{S}}_{\underline{f}} \times \hat{\underline{S}}_{\underline{h}}] ; \hat{\underline{S}}_{\underline{g}} \rangle\rangle = \underline{\underline{G}}^{(2)}_{\underline{f}\,\underline{h}\,\underline{g}} \tag{6.10}$$

$$\langle(\hat{\underline{S}}_{\underline{a}} \cdot \hat{\underline{S}}_{\underline{b}})\rangle = C_{\underline{a}\,\underline{b}} \tag{6.11}$$

$$\langle[\hat{\underline{S}}_{\underline{f}}, \hat{\underline{S}}_{\underline{g}}]_-\rangle = \underline{\underline{I}}^{(1)}_{\underline{f}\,\underline{g}} \tag{6.12}$$

$$\langle\big[[\hat{\underline{S}}_{\underline{f}} \times \hat{\underline{S}}_{\underline{h}}], \hat{\underline{S}}_{\underline{g}}\big]_-\rangle = \underline{\underline{\underline{I}}}^{(2)}_{\underline{f}\,\underline{k}\,\underline{g}} \tag{6.13}$$

where we note that while in equations (6.3) and (6.10) two time arguments occur, in equations (6.11) to (6.13) all time arguments coincide. Equations (6.6) and (6.7) now become

$$\omega\, \underline{\underline{G}}^{(1)}_{\underline{fg}} = \underline{\underline{I}}^{(1)}_{\underline{fg}} + i \sum_{\underline{a}} J_{\underline{fa}}\, \underline{\underline{\underline{G}}}^{(2)}_{\underline{fag}} \quad , \tag{6.14}$$

and

$$\omega\, \underline{\underline{\underline{G}}}^{(2)}_{\underline{fag}} = \underline{\underline{\underline{I}}}^{(2)}_{\underline{fag}} + i \sum_{\underline{b}} \Big\{ J_{\underline{fb}}\Big[C_{\underline{fa}}\underline{\underline{G}}^{(1)}_{\underline{bg}} - C_{\underline{ba}}\underline{\underline{G}}^{(1)}_{\underline{fg}} \Big] + J_{\underline{ab}}\Big[C_{\underline{fb}}\underline{\underline{G}}^{(1)}_{\underline{ag}} - C_{\underline{fa}}\underline{\underline{G}}^{(1)}_{\underline{bg}} \Big] \Big\} . \tag{6.15}$$

As they stand, equations (6.14) and (6.15) can be solved by the usual lattice Fourier transform, based on the translational invariance of the system. Before doing this, we note that we can insert one exact relation into this approximation. In general $(\hat{\underline{S}}_{\underline{a}} \cdot \hat{\underline{S}}_{\underline{a}}) = S(S+1) \equiv \Gamma_0$. Now in equation (6.15), we always have $a \neq f$ because of equation (6.14) and the fact that $J_{\underline{aa}} = 0$; however, the summed index $\underline{b}$ can equal $\underline{a}$ or $\underline{f}$ and thus $C_{\underline{aa}} = C_0$ appears in front of the Green functions. We know exactly that these are multiplied by Γ_0, and since, therefore, C_0 does not appear in the equations, we should replace it by Γ_0. (No additional approximation is introduced here, and in fact failure to include this exact information leads to completely unphysical results, for example, nearest-neighbour correlations proportional to the temperature. This is not surprising, for if Γ_0 is not included, as in equation (6.15), nothing in the equations specifies the magnitude of the spin S.) To preserve the symmetry of the equations, however, we find it convenient to perform the replacement by subtracting the term in C_0 and adding that in Γ_0 separately, as follows

$$\begin{aligned}\omega\, \underline{\underline{\underline{G}}}^{(2)}_{\underline{fag}} = \underline{\underline{\underline{I}}}^{(2)}_{\underline{fag}} &+ i \sum_{\underline{b}} \Big\{ J_{\underline{fb}}\Big[C_{\underline{fa}}\underline{\underline{G}}^{(1)}_{\underline{bg}} - C_{\underline{ba}}\underline{\underline{G}}^{(1)}_{\underline{fg}} \Big] \\ &+ J_{\underline{ab}}\Big[C_{\underline{fb}}\underline{\underline{G}}^{(1)}_{\underline{ag}} - C_{\underline{fa}}\underline{\underline{G}}^{(1)}_{\underline{bg}} \Big] \Big\} + i\big(C_0 - \Gamma_0\big) J_{\underline{fa}} \Big(\underline{\underline{G}}^{(1)}_{\underline{fg}} - \underline{\underline{G}}^{(1)}_{\underline{ag}}\Big)\end{aligned} \tag{6.16}$$

We now calculate the Fourier transforms and write

$$\underline{\underline{G}}^{(1)}_{\underline{\lambda}}\Big(\underline{\underline{I}}^{(1)}_{\underline{\lambda}}, K_{\underline{\lambda}}, J_{\underline{\lambda}}\Big) = \sum_{\underline{a-b}} e^{i(\underline{\lambda}.\underline{a}-\underline{b})}\, \underline{\underline{G}}^{(1)}_{\underline{ab}}\Big(\underline{\underline{I}}^{(1)}_{\underline{ab}}, C_{\underline{ab}}, J_{\underline{ab}}\Big) , \tag{6.17}$$

$$\underline{\underline{G}}^{(2)}_{\underline{\lambda},\underline{\mu}}\left(\underline{\underline{I}}^{(2)}_{\underline{\lambda},\underline{\mu}}\right) = \sum_{\underline{a}-\underline{c},\underline{b}-\underline{c}} e^{i(\underline{\lambda}.\underline{a}-\underline{c})+i(\underline{\mu}.\underline{b}-\underline{c})}\underline{\underline{G}}^{(2)}_{\underline{abc}}\left(\underline{\underline{I}}^{(2)}_{\underline{abc}}\right) \tag{6.18}$$

and assume inversion symmetry of the lattice so that, e.g., $J_{\underline{\lambda}} = J_{-\underline{\lambda}}$.

Equations (6.14) and (6.15) now become

$$\omega \underline{\underline{G}}^{(1)}_{\underline{\lambda}} = \underline{\underline{I}}^{(1)}_{\underline{\lambda}} + \frac{i}{N}\sum_{\underline{\mu}} J_{\underline{\mu}} \underline{\underline{G}}^{(2)}_{\underline{\lambda}-\underline{\mu},\underline{\mu}} \tag{6.19}$$

$$\omega \underline{\underline{G}}^{(2)}_{\underline{\lambda},\underline{\nu}} = \underline{\underline{I}}^{(2)}_{\underline{\lambda},\underline{\nu}} + i\left[(K_{\underline{\nu}}-K_{\underline{\lambda}})J_{\underline{\nu}+\underline{\lambda}} + K_{\underline{\lambda}}J_{\underline{\lambda}} - K_{\underline{\nu}}J_{\underline{\nu}}\right]\underline{\underline{G}}^{(1)}_{\underline{\lambda}+\underline{\nu}}$$

$$+ i\,(\Gamma_o - C_o)(J_{\underline{\lambda}} - J_{\underline{\nu}})\,\underline{\underline{G}}^{(1)}_{\underline{\lambda}+\underline{\nu}} \tag{6.20}$$

Eliminating $\underline{\underline{G}}^{(2)}_{\underline{\lambda},\underline{\nu}}$ and solving for $\underline{\underline{G}}^{(1)}_{\underline{\lambda}}$, we find

$$\underline{\underline{G}}^{(1)}_{\underline{\lambda}} = \frac{\omega \underline{\underline{I}}^{(1)}_{\underline{\lambda}} + (i/N)\sum_{\underline{\mu}} J_{\mu}\, \underline{\underline{I}}^{(2)}_{\underline{\lambda}-\underline{\mu},\underline{\mu}}}{\omega^2 - \omega^2_{\underline{\lambda}}}, \tag{6.21}$$

where

$$\omega^2_{\underline{\lambda}} = \frac{1}{N}\sum_{\underline{\mu}}\left\{K_{\underline{\mu}}(J_{\underline{\mu}} - J_{\underline{\lambda}}) + (\Gamma_o - C_o)J_{\underline{\mu}}\right\}(J_{\underline{\mu}} - J_{\underline{\lambda}-\underline{\mu}}) . \tag{6.22}$$

Equation (6.21) gives us $\underline{\underline{G}}^{(1)}$ but in it occur the $K_{\underline{\mu}}$ as parameters. We must thus express the $K_{\underline{\mu}}$ in terms of $\underline{\underline{G}}^{(1)}$ and then solve the resultant equation — as is done for $\langle \hat{S}^z \rangle$ in the RPA theory.

We first note that the $\underline{\underline{G}}^{(1)}$ are dyadics while the $K_{\underline{\mu}}$ are scalars. We must thus use equations such as

$$\langle(\hat{\underline{S}}_f \cdot \hat{\underline{S}}_g)\rangle = \sum_k \left(\underline{e}_k \cdot \langle\hat{\underline{S}}_f)(\hat{\underline{S}}_g\rangle \cdot \underline{e}_k\right),$$

where the $\underline{e}_k$ $(k = x, y, z)$ are unit vectors along the Cartesian axes and where the dots on the right-hand side of equation (6.22) indicate scalar products with the vector immediately behind or in front of the unit vector.

From the commutation relations for the spin operators one finds after straightforward calculations, that

$$\sum_k \left(\underline{e}_k \cdot \underline{\underline{I}}^{(2)}_{\underline{\lambda},\underline{\nu}} \cdot \underline{e}_k\right) = 2i\,(K_{\underline{\lambda}} - K_{\underline{\nu}}) . \tag{6.24}$$

One can now take equation (6.21), decompose the right-hand side into partial fractions, and use the fact that in the $K_{\underline{\lambda}}$ the time arguments coincide to find the implicit equation

$$K_{\underline{\lambda}} = \frac{\coth\frac{1}{2}\beta\omega_{\underline{\lambda}}}{\omega_{\underline{\lambda}}}\,\frac{1}{N}\sum_{\underline{\mu}} K_{\underline{\mu}}(J_{\underline{\mu}} - J_{\underline{\lambda}-\underline{\mu}}) . \tag{6.25}$$

To make the discussion more transparent, we shall restrict ourselves to the case of nearest-neighbour interactions only, so that, if $\underline{\Delta}$ indicates here and henceforth a nearest-neighbour vector

$$J_{\underline{a}\underline{b}} = J\, \delta_{\underline{a},\underline{b}+\underline{\Delta}} , \tag{6.26}$$

where we, moreover, have assumed J to be independent of $\underline{\Delta}$. We introduce the following notation :

$$\gamma_{\underline{\lambda}} = \frac{1}{z} \sum_{\underline{\Delta}} e^{i(\underline{\lambda}.\underline{\Delta})} , \quad u_{\underline{\lambda}} = 1 - \gamma_{\underline{\lambda}} , \tag{6.27}$$

where z is the number of nearest neighbours. It follows easily that

$$J_{\underline{\lambda}} = z J \gamma_{\underline{\lambda}} . \tag{6.28}$$

We also use the helpful identity

$$\frac{1}{N} \sum_{\underline{\mu}} F_{\underline{\mu}} \gamma_{\underline{\lambda}-\underline{\mu}} = \gamma_{\underline{\lambda}} \frac{1}{N} \sum_{\underline{\mu}} F_{\underline{\mu}} \gamma_{\underline{\mu}} , \tag{6.29}$$

where $F_{\underline{\mu}}$ is the lattice Fourier transform of a function $F_{\underline{r}}$ whose value is independent of lattice direction for $\underline{r} = \underline{\Delta}$. The identity follows from using equation (6.27) for $\gamma_{\underline{\lambda}-\underline{\mu}}$, inverting the transform of $F_{\underline{\mu}}$, using the spherical symmetry of $F_{\underline{\Delta}}$, and retransforming to $F_{\underline{\mu}}$ again.

Equation (6.25) now becomes

$$K_{\underline{\lambda}} = \frac{z J C_1}{\omega_{\underline{\lambda}}} u_{\underline{\lambda}} \coth \tfrac{1}{2} \beta \omega_{\underline{\lambda}} , \tag{6.30}$$

where C_1 is the nearest-neighbour correlation function,

$$C_1 \equiv C_{\underline{a},\underline{a}+\underline{\Delta}} = \langle (\hat{\underline{S}}_{\underline{a}} \cdot \hat{\underline{S}}_{\underline{a}+\underline{\Delta}}) \rangle , \tag{6.31}$$

which is independent of $\underline{\Delta}$ since both the Heisenberg Hamiltonian and the paramagnetic phase are spherically symmetrical.

Similarly, we can manipulate equation (6.22) and find

$$\omega_{\underline{\lambda}}^2 = z^2 J^2 C_1 u_{\underline{\lambda}} (\chi^{-1} + u_{\underline{\lambda}}) , \tag{6.32}$$

where

$$\chi^{-1} = \frac{\Gamma_0 - C_0}{z C_1} + \frac{\sum_{\underline{\Delta}'} C_{\underline{a},\underline{a}+\underline{\Delta}+\underline{\Delta}'} - z C_1}{z C_1} \tag{6.33}$$

Note that, because of the first term, C_0 does not appear in the sum in the second term.

To see the physical meaning of χ we note that for small λ we have $u_{\underline{\lambda}} \approx a\lambda^2$ (we use the assumed inversion symmetry), so that as $\lambda \to 0$ we have

$$K_{\underline{\lambda}} \xrightarrow[\lambda \to 0]{} \frac{2}{\beta z J} \frac{1}{\chi^{-1} + a\lambda^2}, \tag{6.34}$$

the Ornstein-Zernike form for correlation functions near a phase transition. Noting that

$$K_{\underline{\lambda}} = \langle |\hat{\underline{S}}_{\underline{\lambda}}|^2 \rangle, \tag{6.35}$$

and that thermodynamics relates the zero-momentum fluctuations with the static susceptibility χ_{stat} as follows

$$\langle |\hat{\underline{S}}_0|^2 \rangle = \frac{3}{\beta \mu_B^2} \chi_{stat}, \tag{6.36}$$

where μ_B is the Bohr magneton, we see that

$$\chi = \frac{3}{2} \frac{zJ}{\mu_B^2 z} \chi_{stat}, \tag{6.37}$$

where the factor 3 comes from summing over the three spatial directions. Note that because of equation (6.32), χ^{-1} must always be positive for the Green function formalism to go through since $\chi^{-1} < 0$ means that $\omega_{\underline{\lambda}}$ is imaginary for $\underline{\lambda}$ near the origin. Thus the vanishing of χ^{-1}, which equation (6.33) indicates will occur as the temperature is lowered from infinity, if the $C_{\underline{a}\,\underline{b}}$ increase as intuitively expected, defines the breakdown of the decoupling approximation. We may interpret this as due to the onset of ferromagnetic ordering: thus the transition temperature T_c for this theory is that at which χ^{-1} vanishes.

For positive χ^{-1}, the poles of the Green function, i.e., what this theory sees as excitations, occur at the energies $\pm\omega_{\underline{\lambda}}$ given by equation (6.33): $\omega_{\underline{\lambda}}$ as a function of λ has quite different forms for high and low momenta, if χ^{-1} is not too large:

$$\omega_{\underline{\lambda}} \approx zJ\,(C_1\chi^{-1})^{\frac{1}{2}}\, u_{\underline{\lambda}}^{\frac{1}{2}}, \quad \chi^{-1} \gg u_{\underline{\lambda}}\;; \tag{6.38}$$

$$\omega_{\underline{\lambda}} \approx zJ\, C_1^{\frac{1}{2}}\, u_{\underline{\lambda}}, \qquad \chi^{-1} \ll u_{\underline{\lambda}}\,. \tag{6.39}$$

The change-over occurs at momenta of the order of λ_0, which is defined such that $u_{\lambda_0} = \chi^{-1}$. For χ^{-1} very small, since $u_{\underline{\lambda}} \approx a\lambda^2$, we have $\lambda_0 = \sqrt{(\chi^{-1}/a)}$. Note that the high-momentum form has the momentum dependence of spin-waves, which tends to confirm the idea that quasi-spin-waves should appear near the critical point for small but non-zero wave-vector, and that λ_0 is proportional to $\chi^{-\frac{1}{2}}$. In this order, the theory reveals nothing about the damping of such excitations, but it definitely supports the suggestions of their existence. The low-momentum form for $\omega_{\underline{\lambda}}$ is linear in λ for small λ. No such excitation has been suggested so far. It would be interesting to see what the inclusion of damping does to this form.

Using the relation (in fact the definition of C_0, since it is not a decoupling parameter)

$$C_0 = \frac{1}{N} \sum_{\underline{\lambda}} K_{\underline{\lambda}} , \tag{6.40}$$

we can write our self-consistent equations (6.30) and (6.32) in the form

$$C_1 = \frac{1}{N} \sum_{\underline{\lambda}} K_{\underline{\lambda}} \gamma_{\underline{\lambda}} , \tag{6.41}$$

$$C_1 = (\chi^{-1} + 1) = \frac{1}{N} \sum_{\underline{\lambda}} K_{\underline{\lambda}} \left(\gamma_{\underline{\lambda}}^2 - \frac{1}{z} \right) + \frac{\Gamma_0}{z} , \tag{6.42}$$

where $K_{\underline{\lambda}}$ satisfies equation (6.30) and $\omega_{\underline{\lambda}}$ is given by equation (6.32).

We shall not give all the details of the evaluation of equations (6.41) and (6.42), but restrict ourselves to quoting a few results, both near T_c which was defined as the point where χ^{-1} vanishes, and at high temperatures.

First of all, we find for T_c the equation

$$T_c = \frac{2}{3} T_c^{(0)} \left[\frac{3\Gamma_0 + \left(\Gamma_0 k_B J / 3 T_c^{(0)} \right)(z-1) - 1}{3\Gamma_0} \right] , \tag{6.43}$$

where $T_c^{(0)}$ is the RPA transition temperature given by equation (3.52), and Γ_0 is still given by the relation

$$\Gamma_0 = S(S+1) . \tag{6.44}$$

Near, but above T_c, we find that the susceptibility χ and the specific heat C behave as follows:

$$\chi = a(T - T_c)^2 + \ldots , \tag{6.45}$$

$$C = b + c(T - T_c) + \ldots , \tag{6.46}$$

where a, b and c are constants: As in the spherical and in the molecular-field model, χ behaves as $(T - T_c)^2$.

At high temperatures, we find for the susceptibility the expression

$$\chi = \frac{\mu_B^2 \, S(S+1)}{3 k_B T} + \dots \quad , \tag{6.47}$$

which is asymptotically exact. The next term (in T^{-2}) differs from the exact one.

The specific heat at high temperatures is proportional to T^{-2}, but the proportionality factor is not quite the exact one.

A measurement for the fit of our approximation is the value we obtain for the sum C_0 of equation (6.40). In an exact solution, it should equal Γ_0. Near T_c it differs from Γ_0 by an amount which depends on the value of S. If $S = \frac{1}{2}$ it differs at most 10 per cent, the difference decreasing rapidly with increasing S. At high temperatures we find

$$C_0 = \Gamma_0 + \frac{1}{3}\left[\frac{S(S+1)\, z\, J}{4 k_B T}\right]^2 + \dots \quad . \tag{6.48}$$

2.7 Lines' Approach to Green Function Decoupling

We have seen that one of the basic problems in the theory of Green functions is the decoupling. The decoupling is usually performed as early as possible in order to keep the mathematical complexity to a minimum. Essentially we are dealing with an equation of motion which after Fourier transforming with respect to the lattice is of the form (compare equations (3.10) and (3.14), (5.3) and (3.14), or (6.19) and (6.20))

$$G_{\underline{q}}(E) = F + \Delta_{\underline{q}}(E) \quad , \tag{7.1}$$

where $G_{\underline{q}}(E)$ is the Green function we are looking for, and $\Delta_{\underline{q}}(E)$ a higher-order Green function.

The decoupling then consists in finding a relation of the form

$$\Delta_{\underline{q}}(E) = M_{\underline{q}}(E)\; G_{\underline{q}}(E) \quad , \tag{7.2}$$

which leads to the following solution of equation (7.1):

$$G_{\underline{q}}(E) = \frac{F}{E - M_{\underline{q}}(E)} \quad , \tag{7.3}$$

which leads to the spin-wave energies which are solutions of the equation

$$E - M_{\underline{q}}(E) = 0 \quad . \tag{7.4}$$

In fact, the decoupling is done not in equation (7.1) but in the real space equations (equations (3.10), (5.3), or (6.8)), and if one looks at it carefully, as was done by M.E. Lines to whom we refer for further details, one finds that the decoupling used by us in subsection 2.3 corresponds to writing

$$M_{\underline{q}}(E) = \xi(T)\, E_0'(\underline{q}) \quad , \tag{7.5}$$

with $E_0'(\underline{q})$ the simple (unrenormalized) spin-wave energy, given by the equation (compare equation (3.19); we put $B=0$) :

$$E_0'(\underline{q}) = 2[K(0) - K(\underline{q})] \quad , \tag{7.6}$$

and $\xi(T)$ given by the equation

$$\xi(T) = \tfrac{1}{2}\sigma = \tfrac{1}{2}\langle S^z\rangle \quad . \tag{7.7}$$

In more sophisticated decoupling schemes only the form of $\xi(T)$ is different, but <u>it remains independent of $\underline{q}$</u>. We can describe equations (7.4) and (7.5) as giving a spin-wave (or magnon) renormalization which is wave-vector independent. They must thus predict that if there is a phase transition, at the transition temperature <u>all</u> spin-wave-like excitations must have vanishing energies, in contrast to experimental evidence of both neutron diffraction and Raman experiments that short-wavelength excitations persist well into the paramagnetic phase (compare the discussion in subsection 2.6).

Lines has proposed to remove the restriction that $\xi(T)$ should be wave-vector-independent and suggested instead of equation (7.5) the relation

$$M_{\underline{q}}(E) = \xi_{\underline{q}}(T)\, E_0'(\underline{q}) \quad , \tag{7.8}$$

and we find, for instance, for the magnetization again equation (3.22) with $E_{\underline{q}}$ replaced by $M_{\underline{q}}(E)$ from equation (7.8).

We see immediately one of the advantages of this scheme. In subsection 2.3 we found that the temperature dependence of $E_{\underline{q}}$ was given by the incorrect equation (3.59). By a suitable choice of (7.8) this now can be remedied. As a first example Lines has studied a two-dimensional nearly isotropic Heisenberg ferromagnet and used a $\xi_{\underline{q}}(T)$ which is essentially given by (7.7) — generalized for general spin-values — for q much smaller than the inverse of a coherence length L and with $\xi_{\underline{q}}(T)$ equal to essentially the square root of the C_1 of equation (6.31) for large q. One may expect L to become very large at low temperatures, so that, indeed, at low temperatures all magnon energies will be renormalized — and incidentally in such a way as to lead to expression (3.40).

We refer to Lines' paper for details of his calculations and a discussion of his results. For our purposes it suffices to indicate a different way to obtain the Green function we are looking for, involving an appeal to experimental or phenomenological arguments in deciding on the approximation to use.

CHAPTER 3

The Pair Hamiltonian Model of an Imperfect Bose Gas

In this section we consider the pair Hamiltonian model of the imperfect Bose gas, and solve it exactly in the absence of Bose-Einstein condensation both by Green function methods and by diagrammatic techniques. Luban's sufficient condition for Bose-Einstein condensation is shown to be equivalent to the condition that the two-body potential be greater than or equal to zero. The problem for the condensed region is solved exactly by a method related to Luban's original one.

3.1 Introduction

The starting point for most many-body calculations of properties of many-boson systems is the second-quantized form of the Hamiltonian:

$$H_{op} = \sum_{\underline{k}} T_{\underline{k}} \hat{a}^{+}_{\underline{k}} \hat{a}_{\underline{k}} + \frac{1}{2V} \sum_{\underline{\ell},\underline{p},\underline{q}} v(\underline{q}) \hat{a}^{+}_{\underline{p}+\underline{q}} \hat{a}^{+}_{\underline{\ell}-\underline{q}} \hat{a}_{\underline{\ell}} \hat{a}_{\underline{p}} , \qquad (1.1)$$

where V is the volume of the system, $T_{\underline{k}} = \frac{k^2}{2m} - \mu$, m is the mass of a boson, μ the chemical potential, $v(\underline{q})$ is the Fourier transform of the two-body potential, and $\hat{a}^+$ and $\hat{a}$ are boson creation and annihilation operators. The problem posed by the Hamiltonian equation (1.1) is in the general form insoluble, but one can derive from equation (1.1) model Hamiltonians which, one hopes will still reflect some of the properties of the original Hamiltonian while being amenable to analytical studies. The simplest meaningful example is the Bogolyubov Hamiltonian which is obtained from equation (1.1) by omitting all interaction terms containing less than two operators $\hat{a}_0$ and $\hat{a}_0^+$. If the zero-momentum state has a macroscopic occupation number $\langle \hat{a}_0^+ \hat{a}_0 \rangle / V$ is finite. Bogolyubov then replaced the operators $\hat{a}_0$ and $\hat{a}_0^+$ by the c-number $\sqrt{\langle \hat{a}_0^+ \hat{a}_0 \rangle}$, and the resulting Hamiltonian was diagonalized by a canonical transformation. The energy spectrum thus obtained has the correct qualitative shape.

The pair Hamiltonian model we study is a generalization of Bogolyubov's Hamiltonian *, and has the form

---

* See appendix to this section for a discussion of the Bogolyubov Hamiltonian.

$$\hat{H}_p = \sum_{\underline{k}} T_{\underline{k}} \hat{a}^+_{\underline{k}} \hat{a}_{\underline{k}} + \frac{v(0)}{2V} \sum_{k,p} \hat{a}^+_{\underline{k}} \hat{a}^+_{\underline{p}} \hat{a}_{\underline{p}} \hat{a}_{\underline{k}}$$

$$+ \frac{1}{2V} \sum_{\underline{k},\underline{p}(\neq \pm \underline{k})} v(\underline{p}-\underline{k}) \hat{a}^+_{\underline{k}} \hat{a}^+_{\underline{p}} \hat{a}_{\underline{k}} \hat{a}_{\underline{p}} + \frac{1}{2V} \sum_{\underline{k},\underline{p}(\neq \underline{k})} v(\underline{p}-\underline{k}) \hat{a}^+_{\underline{p}} \hat{a}^+_{-\underline{p}} \hat{a}_{-\underline{k}} \hat{a}_{\underline{k}} \ . \tag{1.2}$$

Only forward scattering, exchange scattering and pair scattering terms are retained in the interaction Hamiltonian. Variational techniques and the Bogolyubov approximation have been applied to this problem (Zubarev and Tserkovnikov, 1958; Valatin and Butler, 1958; Girardeau and Arnowitt, 1959; Girardeau, 1962), and Luban (1962), using Wentzel's (1960) method of thermodynamically equivalent Hamiltonians, claimed that the pair Hamiltonian has the same thermodynamic properties as a certain Hamiltonian quadratic in the creation and annihilation operators. This equivalent Hamiltonian was diagonalozed by a canonical transformation and its thermodynamic properties were investigated. Bose-Einstein (B.E.) condensation was shown to occur for a certain class of potentials.

In subsection 3.2 we derive the basic equations of Luban's paper by employing double-time temperature-dependent Green functions, and in the following section we show that our decoupling procedure is exact in the limit of infinite volume, provided there is no B.E. condensation. Luban's sufficient condition for B.E. condensation is simplified in subsection 3.4. Subsection 3.5 is devoted to a diagrammatic solution of the problem above the B.E. transition temperature. Some inadequacies in Luban's solution are indicated in subsection 3.6, before we give in subsection 3.7 an exact solution to the problem using a method close in spirit to Luban's own.

3.2 A Green Function Solution of the Problem

The retarded and advanced Green functions $\langle\langle A_{op}(t);B_{op}(t')\rangle\rangle_r$ and $\langle\langle A_{op}(t);B_{op}(t')\rangle\rangle_a$ are defined by the relations

$$\left.\begin{aligned} \langle\langle A_{op}(t);B_{op}(t'\rangle\rangle_r &= -i\,\theta(t-t')\,\langle [A_{op}(t),B_{op}(t')]_-\rangle \ , \\ \langle\langle A_{op}(t);B_{op}(t')\rangle\rangle_a &= i\,\theta(t'-t)\,\langle [A_{op}(t),B_{op}(t')]_-\rangle \ , \\ \theta(t) &= \begin{cases} 1\ , & t > 0 \ , \\ 0\ , & t < 0 \ . \end{cases} \end{aligned}\right\} \tag{2.1}$$

The time Fourier transforms of the Green functions may be used to define a function analytic in the whole of the complex energy plane, apart from a cut along the real axis. For details of the Green functions we refer to the articles by Zubarev (1960) and ter Haar (1962) (see also Chapter 1 of the present lecture notes).

The equation of motion of the Green function is given by

$$E\langle\langle A_{op};B_{op}\rangle\rangle = \frac{1}{2\pi}\langle [A_{op},B_{op}]_-\rangle + \langle\langle [A_{op},H_{op}]_- ; B_{op}\rangle\rangle \ , \qquad (2.2)$$

where the Green functions in equation (2.2) are the time Fourier transforms of those in equation (2.1).

We now introduce the two Green functions

$$G_{\underline{\ell}} = \langle\langle \hat{a}_{\underline{\ell}};\hat{a}^+_{\underline{\ell}}\rangle\rangle \quad \text{and} \quad \Gamma_{\underline{\ell}} = \langle\langle \hat{a}^+_{-\underline{\ell}};\hat{a}^+_{\underline{\ell}}\rangle\rangle \ . \qquad (2.3)$$

The second of these Green functions is rather interesting since one would expect it to vanish if thermal averages are calculated, using the Hamiltonian $\hat{H}_p$, which conserves the total number of particles. To avoid this difficulty, thermal averages will be interpreted as 'quasi-averages' (Bogolyubov, 1960). A similar situation is encountered in the theory of ferromagnetism. Since the Hamiltonian of an isotropic Heisenberg ferromagnet is invariant under rotation of spins on the lattice sites, one would expect its magnetization to be zero. However, if one introduces a small external field, the magnetization in the direction of the field will have a non-zero expectation value even in the limit of an infinitesimal magnetic field. Similarly, the Green function Γ_{ℓ} need not be zero if we add a term $\lambda\sum_{\underline{k}}(\hat{a}^+_{\underline{k}}\hat{a}^+_{-\underline{k}} + \hat{a}_{-\underline{k}}\hat{a}_{\underline{k}})$ to the Hamiltonian. 'Quasi-average' is the expression used by Bogolyubov to describe averages obtained by modifying the Hamiltonian by the addition of some infinitesimal external field not having the symmetry of the original Hamiltonian. In the work on Green functions which follows, we shall interpret $\langle\ldots\rangle$ as a quasi-average where appropriate.

Using the equation of motion, equation (2.2) and the Hamiltonian $\hat{H}_{\underline{p}}$ given by equation (1.2), we obtain the following equations for $G_{\underline{\ell}}$ and $\Gamma_{\underline{\ell}}$:

$$E\,G_{\underline{\ell}} = \frac{1}{2\pi} + T_{\underline{\ell}}G_{\underline{\ell}} + \frac{v(0)}{2V}\sum \langle\langle \hat{a}^+_{\underline{k}}\hat{a}_{\underline{k}}\hat{a}_{\underline{\ell}};\hat{a}^+_{\underline{\ell}}\rangle\rangle$$

$$+ \frac{1}{V}\sum{}'' v(\underline{\ell}-\underline{k})\langle\langle \hat{a}^+_{\underline{k}}\hat{a}_{\underline{k}}\hat{a}_{\underline{\ell}};\hat{a}^+_{\underline{\ell}}\rangle\rangle + \frac{1}{V}\sum{}'' v(\underline{\ell}-\underline{k})\langle\langle \hat{a}^+_{-\underline{\ell}}\hat{a}_{-\underline{k}}\hat{a}_{\underline{k}};\hat{a}^+_{\underline{\ell}}\rangle\rangle \ , \qquad (2.4a)$$

$$E\,\Gamma_{\underline{\ell}} = -\,T_{\underline{\ell}}\Gamma_{\underline{\ell}} - \frac{v(0)}{2V}\sum \langle\langle \hat{a}^{+}_{-\underline{\ell}}\hat{a}^{+}_{-\underline{k}}\hat{a}_{-\underline{k}}\,;\,\hat{a}^{+}_{\underline{\ell}}\rangle\rangle$$

$$-\frac{1}{V}\sum{}'' v(\underline{\ell}-\underline{k})\,\langle\langle \hat{a}^{+}_{-\underline{\ell}}\hat{a}^{+}_{-\underline{k}}\hat{a}_{-\underline{k}}\,;\,\hat{a}^{+}_{\underline{\ell}}\rangle\rangle - \frac{1}{V}\sum{}' v(\underline{\ell}-\underline{k})\,\langle\langle \hat{a}^{+}_{\underline{k}}\hat{a}^{+}_{-\underline{k}}\hat{a}_{\underline{\ell}}\,;\,\hat{a}^{+}_{\underline{\ell}}\rangle\rangle \quad , \tag{2.4b}$$

where Σ, Σ', and Σ'' are abbreviations for $\sum_{\underline{k}}$, $\sum_{\underline{k}(\neq\underline{\ell})}$ and $\sum_{\underline{k}(=\pm\underline{\ell})}$, respectively. We decouple the equations of motion, equations (2.4) by factorizing the higher-order Green functions, putting

$$\begin{aligned}
\langle\langle \hat{a}^{+}_{\underline{k}}\hat{a}^{+}_{-\underline{k}}\hat{a}_{\underline{\ell}}\,;\,\hat{a}^{+}_{\underline{\ell}}\rangle\rangle &= A_{\underline{k}}\,G_{\underline{\ell}} \quad ,\\
\langle\langle \hat{a}^{+}_{\underline{k}}\hat{a}_{k}\hat{a}_{\underline{\ell}}\,;\,\hat{a}^{+}_{\underline{\ell}}\rangle\rangle &= n_{\underline{k}}\,G_{\underline{\ell}} \quad ,\\
\langle\langle \hat{a}^{+}_{\underline{k}}\hat{a}_{\underline{k}}\hat{a}^{+}_{-\underline{\ell}}\,;\,\hat{a}^{+}_{\underline{\ell}}\rangle\rangle &= n_{\underline{k}}\,\Gamma_{\underline{\ell}} \quad ,\\
\langle\langle \hat{a}_{-\underline{k}}\hat{a}_{\underline{k}}\hat{a}^{+}_{-\ell}\,;\,\hat{a}^{+}_{\ell}\rangle\rangle &= A_{\underline{k}}\,\Gamma_{\ell} \quad ,
\end{aligned} \tag{2.5}$$

where

$$\begin{aligned}
n_{\underline{k}} &= \langle \hat{a}^{+}_{\underline{k}}\hat{a}_{\underline{k}}\rangle \quad ,\\
A_{\underline{k}} &= \langle \hat{a}^{+}_{\underline{k}}\hat{a}^{+}_{-\underline{k}}\rangle = \langle \hat{a}_{\underline{k}}\hat{a}_{-\underline{k}}\rangle \quad .
\end{aligned} \tag{2.6}$$

Substitution of the factorization equations (2.5) into equation (2.4) gives

$$\begin{aligned}
E\,G_{\underline{\ell}} &= \frac{1}{2\pi} + \left\{T_{\underline{\ell}} + \frac{Nv(0)}{V} + \frac{1}{V}\sum{}'' v(\underline{\ell}-\underline{k})\,n_{\underline{k}}\right\}G_{\underline{\ell}} + \left\{\frac{1}{V}\sum{}' v(\underline{\ell}-\underline{k})\,A_{\underline{k}}\right\}\Gamma_{\underline{\ell}} \quad ,\\
E\,\Gamma_{\underline{\ell}} &= -\left\{T_{\underline{\ell}} + \frac{Nv(0)}{V} + \frac{1}{V}\sum{}'' v(\underline{\ell}-\underline{k})\,n_{\underline{k}}\right\}\Gamma_{\underline{\ell}} - \left\{\frac{1}{V}\sum{}' v(\underline{\ell}-\underline{k})\,A_{\underline{k}}\right\}G_{\underline{\ell}} \quad ,
\end{aligned} \tag{2.7}$$

where $N = \Sigma\, n_{\underline{k}}$.

If we define

$$\begin{aligned}
\tilde{\epsilon}_{\underline{\ell}} &= T_{\underline{\ell}} + \frac{N}{V}\,v(0) + \frac{1}{V}\sum{}'' v(\underline{\ell}-\underline{k})\,n_{\underline{k}} \quad ,\\
\Delta_{\underline{\ell}} &= \frac{1}{V}\sum{}' v(\underline{\ell}-\underline{k})\,A_{\underline{k}} \quad ,
\end{aligned} \tag{2.8}$$

the coupled equations (2.7) now read

$$\begin{aligned}
(E-\tilde{\epsilon}_{\underline{\ell}})\,G_{\underline{\ell}} - \Delta_{\underline{\ell}}\Gamma_{\underline{\ell}} &= \frac{1}{2\pi} \quad ,\\
(E+\tilde{\epsilon}_{\underline{\ell}})\,\Gamma_{\underline{\ell}} + \Delta_{\underline{\ell}}\Gamma_{\underline{\ell}} &= 0 \quad .
\end{aligned} \tag{2.9}$$

Solving the equations (2.9) for $G_{\underline{\ell}}$ and $\Gamma_{\underline{\ell}}$ gives

$$G_{\underline{\ell}} = \frac{1}{2\pi} \frac{E + \tilde{\epsilon}_{\underline{\ell}}}{E^2 - \tilde{\epsilon}^2_{\underline{\ell}} + \Delta^2_{\underline{\ell}}} ,$$

$$\Gamma_{\underline{\ell}} = \frac{-1}{2\pi} \frac{\Delta_{\underline{\ell}}}{E^2 - \tilde{\epsilon}^2_{\underline{\ell}} + \Delta^2_{\underline{\ell}}} . \qquad (2.10)$$

By using the equations relating correlation functions and Green functions, we may find equations for $\tilde{\epsilon}_{\underline{\ell}}$ and $\Delta_{\underline{\ell}}$. Using the results of section 2.1, we see that

$$n_{\underline{\ell}} = \int_{-\infty}^{+\infty} \frac{i\, d\omega}{e^{\beta\omega} - 1} \lim_{\epsilon\to 0} \left[G_{\underline{\ell}}(\omega + i\epsilon) - G_{\underline{\ell}}(\omega - i\epsilon) \right] ,$$

$$A_{\underline{\ell}} = \int_{-\infty}^{+\infty} \frac{i\, d\omega}{e^{\beta\omega} - 1} \lim_{\epsilon\to 0} \left[\Gamma_{\underline{\ell}}(\omega + i\epsilon) - \Gamma_{\underline{\ell}}(\omega - i\epsilon) \right] , \qquad (2.11)$$

where $\beta = 1/k_B T$ (k_B : Boltzmann's constant, T : absolute temperature). Substituting equation (2.10) into equation (2.11), we get

$$n_{\underline{\ell}} = \frac{1}{2}\left[\frac{\tilde{\epsilon}_{\underline{\ell}}}{\epsilon_{\underline{\ell}}} \coth \tfrac{1}{2} \beta \epsilon_{\underline{\ell}} - 1 \right] ,$$

$$A_{\underline{\ell}} = -\frac{1}{2} \frac{\Delta_{\underline{\ell}}}{\epsilon_{\underline{\ell}}} \coth \tfrac{1}{2} \beta \epsilon_{\underline{\ell}} , \qquad (2.12)$$

where

$$\epsilon^2_{\underline{\ell}} = \tilde{\epsilon}^2_{\underline{\ell}} - \Delta^2_{\underline{\ell}} . \qquad (2.13)$$

By using equation (2.12), equation (2.8) become a set of self-consistent equations for $\tilde{\epsilon}_{\underline{\ell}}$ and $\Delta_{\underline{\ell}}$:

$$\tilde{\epsilon}_{\underline{\ell}} = T_{\underline{\ell}} + \frac{N}{V} v(0) + \frac{1}{2V} \sum{}'' v(\underline{\ell} - \underline{k}) \left[\frac{\tilde{\epsilon}_{\underline{k}}}{\epsilon_{\underline{k}}} \coth \tfrac{1}{2} \beta \epsilon_{\underline{k}} - 1 \right] ,$$

$$\Delta_{\underline{\ell}} = -\frac{1}{2V} \sum{}' v(\underline{\ell} - \underline{k}) \frac{\Delta_{\underline{k}}}{\epsilon_{\underline{k}}} \coth \tfrac{1}{2} \beta \epsilon_{\underline{k}} . \qquad (2.14)$$

Our equations (2.14) are identical with the result of using Luban's equations (8) , (9) , (25) and (26) to obtain coupled equations for his functions $f(\underline{\ell})$ and $h(\underline{\ell})$ (our $\tilde{\epsilon}_{\underline{\ell}}$ and $\Delta_{\underline{\ell}}$, respectively).

3.3 Proof of the Asymptotic Exactness of the Solution

In this section we extend to the pair Hamiltonian model the method applied by Bogolyubov, Zubarev and Tserkovnikov (1960) to the reduced Hamiltonian in the theory of superconductivity. We show that in the infinite volume limit the chain of equations of motions of Green functions generated from $G_{\underline{\ell}}$ and $\Gamma_{\underline{\ell}}$ is satisfied exactly, apart from terms of order $1/V$, if we decouple the equations by a generalization of the method given by equations (2.5) for the two-particle Green functions.

The only Green functions of interest may be expressed in one of the forms $\langle\langle \hat{a}^+_{-\underline{\ell}}\hat{Q} ; \hat{a}^+_{\underline{\ell}}\rangle\rangle$ or $\langle\langle \hat{M}\hat{a}_{\underline{\ell}} ; \hat{a}^+_{\underline{\ell}}\rangle\rangle$, where $\hat{M}$ and $\hat{Q}$ are products of factors of the type $\hat{a}_{\underline{k}}\hat{a}^+_{\underline{k}}$, $\hat{a}^+_{-\underline{k}}\hat{a}^+_{\underline{k}}$ or $\hat{a}_{\underline{k}}\hat{a}_{-\underline{k}}$. This is a consequence of the special form of the pair Hamiltonian equation (1.2). The equation of motion, equation (2.2) may be applied to these Green functions and gives

$$E\,\langle\langle \hat{M}\hat{a}_{\underline{\ell}} ; \hat{a}^+_{\underline{\ell}}\rangle\rangle = \frac{1}{2\pi}\langle [\hat{M}\hat{a}_{\underline{\ell}} ; \hat{a}^+_{\underline{\ell}}]_-\rangle + T_{\underline{\ell}}\langle\langle \hat{M}\hat{a}_{\underline{\ell}} ; \hat{a}^+_{\underline{\ell}}\rangle\rangle$$
$$+ \frac{v(0)}{V}\sum \langle\langle \hat{M}\hat{a}^+_{\underline{k}}\hat{a}_{\underline{k}}\hat{a}_{\underline{\ell}} ; \hat{a}^+_{\underline{\ell}}\rangle\rangle + \frac{1}{V}\sum{}'' v(\underline{\ell}-\underline{k})\,\langle\langle \hat{M}\hat{a}^+_{\underline{k}}\hat{a}_{\underline{k}}\hat{a}_{\underline{\ell}} ; \hat{a}^+_{\underline{\ell}}\rangle\rangle$$
$$+ \frac{1}{V}\sum{}' v(\underline{\ell}-\underline{k})\,\langle\langle \hat{M}\hat{a}^+_{-\underline{\ell}}\hat{a}_{\underline{k}}\hat{a}_{-\underline{k}} ; \hat{a}^+_{\underline{\ell}}\rangle\rangle + \langle\langle [\hat{M},\hat{H}_p]_-\,\hat{a}_{\underline{\ell}} ; \hat{a}^+_{\underline{\ell}}\rangle\rangle ,$$

$$E\,\langle\langle \hat{a}^+_{-\underline{\ell}}\hat{Q} ; \hat{a}^+_{\underline{\ell}}\rangle\rangle = \frac{1}{2\pi}\langle [\hat{a}^+_{-\underline{\ell}}\hat{Q} , \hat{a}^+_{\underline{\ell}}]_-\rangle - T_{\underline{\ell}}\langle\langle \hat{a}^+_{-\underline{\ell}}\hat{Q} ; \hat{a}^+_{\underline{\ell}}\rangle\rangle$$
$$- \frac{v(0)}{V}\sum \langle\langle \hat{a}^+_{-\underline{\ell}}\hat{a}^+_{-\underline{k}}\hat{a}_{-\underline{k}}\hat{Q} : \hat{a}^+_{\underline{\ell}}\rangle\rangle - \frac{1}{V}\sum{}'' v(\underline{\ell}-\underline{k})\langle\langle \hat{a}^+_{-\underline{\ell}}\hat{a}^+_{-\underline{k}}\hat{a}_{-\underline{k}}\hat{Q} ; \hat{a}^+_{\underline{\ell}}\rangle\rangle$$
$$- \frac{1}{V}\sum{}' v(\underline{\ell}-\underline{k})\langle\langle \hat{a}^+_{\underline{k}}\hat{a}^+_{-\underline{k}}\hat{a}_{\underline{\ell}}\hat{Q} ; \hat{a}^+_{\underline{\ell}}\rangle\rangle + \langle\langle \hat{a}^+_{-\underline{\ell}}[\hat{Q} , \hat{H}_p]_- ; \hat{a}^+_{\underline{\ell}}\rangle\rangle . \qquad (3.1)$$

In the discussion that follows we shall use a single momentum $\underline{p}$ to denote the pair of momenta $\pm\underline{p}$. First of all let us consider operators $\hat{M}$ and $\hat{Q}$ which contain no operators of momentum $\underline{\ell}$, and which also contain no more than two operators of the same momentum. On the right-hand-side of the equations of motion equation (3.1) we may neglect terms where $\underline{k}$ is equal to $\underline{\ell}$ or to one of the momenta involved in $\hat{M}$ or $\hat{Q}$, provided such Green functions depend on the same power of the volume as the Green function whose equation of motion we are considering. The errors introduced by this approximation will be of order $1/V$ since the interaction Hamiltonian contains a factor $1/V$. However, if there is B.E.-condensation in the zero-momentum state, the Green function $\langle\langle \hat{a}_0^+ \hat{a}_0 \hat{a}_0 ; \hat{a}_0^+ \rangle\rangle$ which occurs in the equation of motion for G_0 will be of the order V^2. G_0 itself is of order V, and, therefore, omission of $\langle\langle \hat{a}_0^+ \hat{a}_0 \hat{a}_0 ; \hat{a}_0^+ \rangle\rangle$ in the equation of motion will lead to errors of order V^0. In the rest of this subsection we restrict our attention to the region where there is no B.E.-condensation. No Green functions in the chain of equations of motion generated from $G_{\underline{\ell}}$ and $\Gamma_{\underline{\ell}}$ contain more than two operators of a given momentum, if we consistently neglect such Green functions in the interaction terms on the right-hand side of the equations.

We now demonstrate that the whole chain of Green functions may be satisfied to within terms of order $1/V$, provided thermal averages are calculated using a different Hamiltonian. The one we choose is Luban's approximating Hamiltonian $\hat{H}^0$, which he claims has the same thermodynamic properties as $\hat{H}_p$. To establish the exactness of Luban's solution above the transition temperature, using the Hamiltonian $\hat{H}_p$ in thermal averages, it is easier to use diagrammatic methods which we describe in section 4.5. $\hat{H}^0$ is given by the relation

$$\hat{H}^0 = U + \sum \left[\tilde{\epsilon}_{\underline{k}} \hat{a}_{\underline{k}}^+ \hat{a}_{\underline{k}} + \tfrac{1}{2} \Delta_{\underline{k}} (\hat{a}_{\underline{k}}^+ \hat{a}_{-\underline{k}}^+ + \hat{a}_{\underline{k}} \hat{a}_{-\underline{k}}) \right] , \tag{3.2}$$

where U is a c-number. As we see, this Hamiltonian couples only states of equal and opposite momenta, and, therefore, thermal averages may be expressed as a product of independent factors, one for each momenta. Thus, for example

$$\langle\langle \hat{M} \hat{a}_{\underline{\ell}} ; \hat{a}_{\underline{\ell}}^+ \rangle\rangle^0 = \langle \hat{M} \rangle^0 \langle\langle \hat{a}_{\underline{\ell}} ; \hat{a}_{\underline{\ell}}^+ \rangle\rangle^0 ,$$

where the superscript o denotes an average using the Hamiltonian H^0.

If the final terms in the equation of motion (3.1) can be neglected, the equations reduce to the equations of motion for $G_{\underline{\ell}}$ and $\Gamma_{\underline{\ell}}$ multiplied by $\langle \hat{M} \rangle^0$ and $\langle \hat{Q} \rangle^0$, respectively. We now show that these terms do in fact vanish.

The commutator $[\hat{M}, \hat{H}_p]_-$ may be expressed as a sum of terms with factors of the type $\hat{a}_{\underline{k}}^+ \hat{a}_{\underline{k}}$, $\hat{a}_{\underline{k}}^+ \hat{a}_{-\underline{k}}^+$, $\hat{a}_{\underline{k}} \hat{a}_{-\underline{k}}$ and their commutators with $\hat{H}_p$. Using the equation (1.2) for $\hat{H}_p$, we find

$$\begin{aligned}
\left[\hat{a}_{\underline{k}} \hat{a}_{-\underline{k}}, \hat{H}_p \right]_- &= 2\, T_{\underline{k}} \hat{a}_{\underline{k}} \hat{a}_{-\underline{k}} + \frac{v(0)}{V} \sum (\hat{a}_{\underline{p}}^+ \hat{a}_{\underline{p}} \hat{a}_{\underline{k}} \hat{a}_{-\underline{k}} + \hat{a}_{\underline{k}} \hat{a}_{-\underline{p}}^+ \hat{a}_{-\underline{p}} \hat{a}_{-\underline{k}}) \\
&+ \frac{1}{V} \sum{}'' v(\underline{p} - \underline{k}) (\hat{a}_{\underline{p}}^+ \hat{a}_{\underline{p}} \hat{a}_{\underline{k}} \hat{a}_{-\underline{k}} + \hat{a}_{\underline{k}} \hat{a}_{-\underline{p}}^+ \hat{a}_{-\underline{p}} \hat{a}_{-\underline{k}}) \\
&+ \frac{1}{V} \sum{}' v(\underline{p} - \underline{k}) (\hat{a}_{-\underline{k}}^+ \hat{a}_{\underline{p}} \hat{a}_{-\underline{p}} \hat{a}_{-\underline{k}} + \hat{a}_{\underline{k}} \hat{a}_{\underline{k}}^+ \hat{a}_{\underline{p}} \hat{a}_{-\underline{p}}) .
\end{aligned} \tag{3.3}$$

Since momenta other than $\pm \underline{k}$ occur only in the interaction term in equation (3.3), terms in the expansion of $\langle\langle [\hat{M},\hat{H}_{\underline{p}}]_- \hat{a}_{\underline{\ell}} ; \hat{a}^+_{\underline{\ell}} \rangle\rangle^o$ may be factorized if some terms of order $1/V$ are neglected. As a result of this factorization, each term in the expansion of $\langle\langle [\hat{M},\hat{H}_{\underline{p}}]_- \hat{a}_{\underline{\ell}} ; \hat{a}^+_{\underline{\ell}} \rangle\rangle^o$ has one factor of the type

$$\langle [\hat{a}_{\underline{k}} \hat{a}_{-\underline{k}}, \hat{H}_p]_- \rangle^o \ , \ \langle [\hat{a}^+_{\underline{k}} \hat{a}^+_{-\underline{k}}, \hat{H}_p]_- \rangle^o \ , \text{ or } \ \langle \hat{a}^+_{\underline{k}} \hat{a}_{\underline{k}}, \hat{H}_p]_- \rangle^o \ .$$

The thermal average of equation (3.3) is

$$\langle [\hat{a}_{\underline{k}} \hat{a}_{-\underline{k}}, \hat{H}_p]_- \rangle^o = 2\left\{ T_{\underline{k}} + \frac{Nv(0)}{V} + \frac{1}{V} \sum{}'' \, v(\underline{p}-\underline{k})\, n^o_{\underline{p}} \right\} A^o_{\underline{k}} + \left\{ \frac{1}{V} \sum{}' \, v(\underline{p}-\underline{k})\, A^o_{\underline{p}} \right\} (1 + 2n^o_{\underline{k}}) \ . \qquad (3.4)$$

The correlation functions $n^o_{\underline{k}}$ and $A^o_{\underline{k}}$ may be found by Green function techniques or by application of a Bogolyubov transformation, and they turn out to be equal to $n_{\underline{k}}$ and $A_{\underline{k}}$. Application of equation (2.12) gives

$$1 + 2n^o_{\underline{k}} = 1 + 2n_{\underline{k}} = \frac{\tilde{\varepsilon}_{\underline{k}}}{\varepsilon_{\underline{k}}} \coth \tfrac{1}{2} \beta \varepsilon_{\underline{k}} \ ,$$

$$A^o_{\underline{k}} = A_{\underline{k}} = -\frac{\Delta_{\underline{k}}}{\varepsilon_{\underline{k}}} \coth \tfrac{1}{2} \beta \varepsilon_{\underline{k}} \ . \qquad (3.5)$$

Substituting equation (3.5) into equation (3.4) and using the definition (2.8) of $\tilde{\varepsilon}_{\underline{k}}$ and $\Delta_{\underline{k}}$ we obtain

$$\langle [\hat{a}_{\underline{k}} \hat{a}_{-\underline{k}}, \hat{H}_p]_- \rangle^o = 0 \ . \qquad (3.6)$$

Similarly

$$\langle [\hat{a}^+_{\underline{k}} \hat{a}_{\underline{k}}, \hat{H}_p]_- \rangle^o = \langle [\hat{a}^+_{\underline{k}} \hat{a}^+_{-\underline{k}}, \hat{H}_p]_- \rangle^o = 0 \ . \qquad (3.7)$$

Thus, since each term in the expansion of $\langle\langle [\hat{M}, \hat{H}_p]_- \hat{a}_{\underline{\ell}} ; \hat{a}^+_{\underline{\ell}} \rangle\rangle^o$ contains a factor similar to those in equations (3.6) and (3.7) we see that

$$\langle\langle [\hat{M}, \hat{H}_p]_- \hat{a}_{\underline{\ell}} ; \hat{a}^+_{\underline{\ell}} \rangle\rangle^o = 0 \ . \qquad (3.8)$$

By related methods it is seen that

$$\langle\langle [\hat{a}^+_{-\underline{\ell}} \hat{Q}, \hat{H}_p]_- ; \hat{a}^+_{\underline{\ell}} \rangle\rangle^o = 0 \ , \qquad (3.9)$$

and the equations of motion (3.1), therefore, reduce to those for $G_{\underline{\ell}}$ and $\Gamma_{\underline{\ell}}$ multiplied by $\langle \hat{M} \rangle^o$ and $\langle \hat{Q} \rangle^o$.

In summary, the whole chain of equations of motion of Green functions generated from $G_{\underline{\ell}}$ and $\Gamma_{\underline{\ell}}$ has been shown to be satisfied to within terms of order $1/V$ by our decoupling procedure, if averages are evaluated using the Hamiltonian $\hat{H}^o$. Our proof is valid only if there is no B.E.-condensation.

In the presence of B.E.-condensation the Green function $\langle\langle \hat{a}_0^+ \hat{a}_0 \hat{a}_0 ; \hat{a}_0^+ \rangle\rangle$ in the equation of motion may not be neglected. However, a consistent decoupling procedure may be found provided it gives correctly the leading-order term in the expansion of $\langle\langle \hat{a}_0^+ \hat{a}_0 \hat{a}_0 ; \hat{a}_0^+ \rangle\rangle$ in powers of V. One such procedure is the Bogolyubov approximation, which neglects fluctuations in the zero-momentum state; although this provides a consistent decoupling method, it is not necessarily exact. For the pair Hamiltonian model it is exact, and this will be proved in subsection 3.7.

3.4 The Condition for Bose-Einstein Condensation

If we assume that none of the quantities in the second of equations (2.12) is singular, we may replace summations by integrations:

$$\sum_{\underline{k}} \rightarrow V \int d^3\underline{k} \quad . \tag{4.1}$$

The second of equations (2.12) now reads

$$\Delta_{\underline{\ell}} = -\frac{1}{2} \int v(\underline{\ell}-\underline{k}) \frac{\Delta_{\underline{k}}}{\epsilon_{\underline{k}}} \coth \tfrac{1}{2} \beta \epsilon_{\underline{k}} \, d^3\underline{k} \; . \tag{4.2}$$

Multiplying both sides of equation (4.2) by $\frac{\Delta_{\underline{\ell}}}{\epsilon_{\underline{\ell}}} \coth \frac{1}{2} \beta \epsilon_{\underline{\ell}} \equiv g_{\underline{\ell}}$ and integrating over $\underline{\ell}$ we find

$$2 \int \frac{\Delta_{\underline{\ell}}^2}{\epsilon_{\underline{\ell}}} \coth \tfrac{1}{2} \beta \epsilon_{\underline{\ell}} \, d^3\underline{\ell} = - \int v(\underline{\ell}-\underline{k}) \, g_{\underline{\ell}} \, g_{\underline{k}} \, d^3\underline{k} \, d^3\underline{\ell} \quad . \tag{4.3}$$

If $\Delta_{\underline{\ell}}$ is not identically equal to zero for all $\underline{\ell}$, the left-hand side of equation (4.3) is positive definite, and there will be an inconsistency unless the right-hand side is also positive definite. Transformation of the right-hand side to coordinate space gives

$$- \int v(\underline{r}) \, g^2(\underline{r}) \, d^3\underline{r} \quad , \tag{4.4}$$

which is negative definite if $v(\underline{r})$ is everywhere greater than or equal to zero. If $v(\underline{r})$ is positive, it is inconsistent to replace summations by integrations. Luban demonstrates that the trivial solution $\Delta_{\underline{\ell}} = 0$ is impossible for certain temperatures, and shows that B.E.-condensation must occur. Thus we conclude that a sufficient condition for B.E.-condensation is

$$v(\underline{r}) \geq 0 \quad . \tag{4.5}$$

This result was probably obscured in Luban's work because he used an angular average of the potential rather than the potential itself, which we used.

3.5 A Graphical Solution of the Problem

It is well known that considerable difficulties are encountered when one attempts to apply the usual graphical perturbation theory to systems of bosons with B.E.-condensation. In this section we, therefore, confine our attention to the pair Hamiltonian model without condensation.

We use a generalization to the many-boson problem of the formalism of Abrikosov, Gor'kov and Dzyaloshinskii (1965). In future we shall abbreviate this reference to AGD. AGD discuss the linked-cluster expansion for Fermi systems, although their proof of Wick's theorem is incorrect. Their results are easily generalized to Bose systems without condensation. As well as normal contractions of the type $\langle \hat{a}^+_{\underline{k}} \hat{a}_{\underline{k}} \rangle$ which are non-zero in the unperturbed system, we introduce anomalous ones like $\langle \hat{a}_{\underline{k}} \hat{a}_{-\underline{k}} \rangle$ which vanish identically if averages are evaluated using the unperturbed Hamiltonian, but which may have a finite value if $\langle \ldots \rangle$ is interpreted as a quasi-average. The diagrammatic expansion will involve lines representing the anomalous correlations as well as those for normal correlations. The anomalous correlations cannot be determined by any simple perturbation expansion, but must be found in a self-consistent way from a homogeneous integral equation. This is analogous to the procedure described for fermions.

The perturbation expansion may be resummed to give results similar to those of Hugenholtz and Pines (1959) for the Bose gas at zero temperature. We denote the irreducible self-energy part with a single incoming and a single outgoing line by $\Sigma_{11}(\underline{p}, \omega_n)$, and those with two outgoing or two incoming lines, by $\Sigma_{20}(\underline{p}, \omega_n)$ and $\Sigma_{02}(\underline{p}, \omega_n)$. We use the symbol $G(\underline{p}, \omega_n)$ for the single-particle Green function, and $G_n(\underline{p}, \omega_n) = [G_0^{-1}(\underline{p}, \omega_n) - \Sigma_{11}(\underline{p}, \omega_n)]^{-1}$ for the single-particle Green function consisting of a chain of Σ_{11} insertions, but with no Σ_{02} insertions. The function $G_0(\underline{p}, \omega_n) = [\omega_n - T_{\underline{p}}]^{-1}$ is the unperturbed Green function. Finally we define the irreducible vertex function Γ. This consists of the sum of all linked graphs with two incoming lines and two outgoing lines, excluding graphs where the part of the graph containing the incoming lines may be separated from the rest of the graph by cutting two internal propagators.

The graphs for Σ_{11} and Σ_{02} have the same form for the Hamiltonian H_{op} given by equation (1.1) and $\hat{H}_p$ given by equation (1.2). One important difference is that a large number of graphs give zero contribution in the infinite-volume limit, if the interaction matrix element for scattering of two particles of momenta $\underline{p}_1$ and $\underline{p}_2$ into states of momenta $\underline{p}_3$ and $\underline{p}_4$ is

$$\frac{1}{V}\, v(\underline{p}_1 - \underline{p}_3)\ \delta(\underline{p}_1 + \underline{p}_2 - \underline{p}_3 - \underline{p}_4)$$

whereas the corresponding expression for the pair Hamiltonian is

$$\frac{1}{V}\, v(\underline{p}_1 - \underline{p}_3)\ \delta(\underline{p}_1 + \underline{p}_2 - \underline{p}_3 - \underline{p}_4)\left[\delta(\underline{p}_1 - \underline{p}_4) + \delta(\underline{p}_1 - \underline{p}_3) + \delta(\underline{p}_1 - \underline{p}_2)\right]. \qquad (5.2)$$

Strictly speaking, equation (5.2) should contain terms like $-\delta(\underline{p}_1 - \underline{p}_4)\ \delta(\underline{p}_1 - \underline{p}_3)$ inside the square brackets, but we shall not write these out explicitly since they do not alter our arguments. In the early stages of the discussion we shall use a discrete spectrum for the energy levels of the bosons, and thus the delta-functions must be interpreted as Kronecker symbols: only the final independent summations over momentum space will be replaced by integrations. This enables us to avoid difficulties associated with the integrals of products of Dirac delta functions with the same arguments.

We now look at the volume dependence of terms in the expansion of Σ_{11} in terms of renormalized propagators. Each term gives a volume-independent contribution if the full Hamiltonian (5.1) is used, but when the pair-Hamiltonian (5.2) is used, many terms will vanish in the infinite-volume limit because the extra delta-functions in the square brackets restrict the number of independent summations over momentum space.

Some terms will still produce volume-independent contributions; two examples are the Hartree and Hartree-Fock graphs shown in Fig. 3.1. In these two examples momentum conservation automatically ensures that $\underline{p}_1 = \underline{p}_3$ or $\underline{p}_1 = \underline{p}_4$, and therefore, the delta-functions in the square bracket in equation (5.2) produce no further restriction on the momentum-space summation. On the other hand, the graph in Fig. 3.2 will give a contribution of order $1/V$ since there are two interactions each giving a factor V. We conclude that graphs may be neglected unless momentum conservation automatically ensures that all scattering processes are either forward scattering $(\underline{p}_1 = \underline{p}_3)$, exchange scattering $(\underline{p}_1 = \underline{p}_4)$ or pair scattering $(\underline{p}_1 = -\underline{p}_2)$.

Fig. 3.1

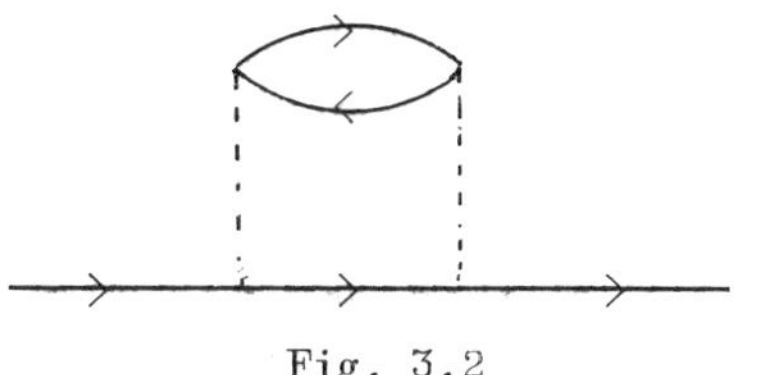

Fig. 3.2

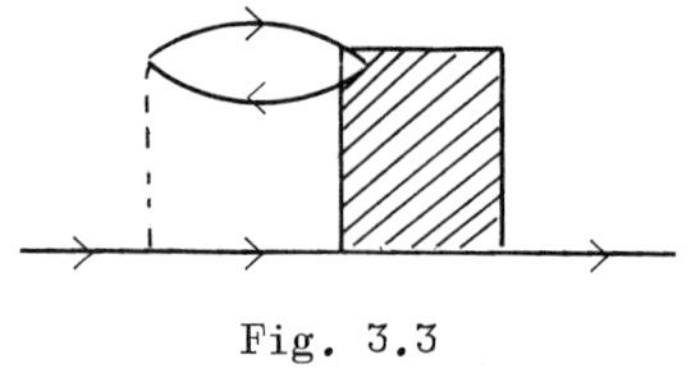

Fig. 3.3

Let us now demonstrate that the Hartree and Hartree-Fock diagrams are the only ones that need to be considered. Higher-order graphs may be represented schematically by the diagram shown in Fig. 3.3, where the shaded box represents a graph for the vertex function. A linked graph with 2n propagators with free ends gives a contribution of order $1/V^{n-1}$, if the full Hamiltonian is used, and cannot produce a larger contribution if the interaction is given by the pair Hamiltonian. The volume-dependence of a graph is easily calculated, since it is the number of independent loops in the diagram less the number of interactions. Thus the contribution from Fig. 3.3 has two factors $1/V$, one each for the interaction and the vertex function, and a single momentum-space summation. If no propagators are proportional to the volume, a condition we take to be equivalent to the absence of B.E.-condensation, the graph contributes a term of order $1/V$ which may be neglected. Thus we conclude that the only graphs contributing to Σ_{11} are the Hartree and Hartree-Fock ones shown in Fig. 3.1.

The integral equation for Σ_{02} has the graphical form shown in Fig. 3.4.

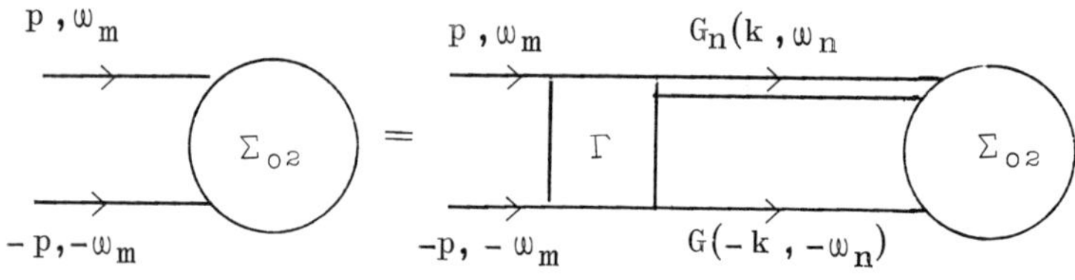

Fig. 3.4

This equation has an asymmetrical appearance, but this is essential to avoid over-summing diagrams. To simplify this equation we show that Γ is effectively the bare interaction, since all the other contributions vanish in the infinite volume limit. Higher-order terms may be expressed symbolically by the graphs in Fig. 3.5, where the shaded boxes are again vertex functions.

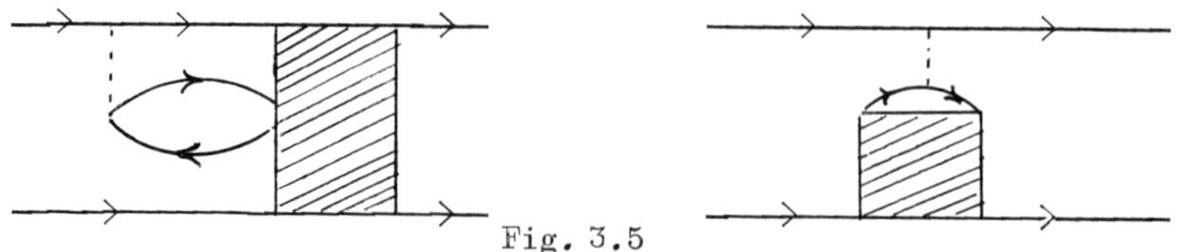

Fig. 3.5

Using the result on the size of vertex functions, we see that the graphs produce contributions of order $1/V^2$ at most. If the internal lines refer to the single-particle Green function containing two creation operators or two annihilation operators, the result will remain true. Consequently the irreducible vertex function Γ may be replaced by the bare interaction; this means that Σ_{02} is independent of frequency, since the bare interaction does not depend on frequency.

Application of the rules of thermodynamic perturbation theory to the calculation of the graphs in Fig. 3.1, which we have shown are the only ones contributing to Σ_{11}, gives

$$\Sigma_{11}(\underline{p}) = -\frac{v(0)}{\beta V}\sum_{\underline{k},\omega_n} G(\underline{k},\omega_n)\, e^{\omega_n \tau} - \frac{1}{\beta V}\sum_{\underline{k},\omega_n} v(\underline{p}-\underline{k})\, G(\underline{k},\omega_n) e^{\omega_n \tau}, \quad (5.3)$$

where $\tau \to +0$. The graphical equation for Σ_{02}, given in Fig. 3.4, is expressed analytically as

$$\Sigma_{02}(\underline{p}) = \frac{1}{\beta V}\sum_{\underline{k},\omega_n} v(\underline{p}-\underline{k})\; G_n(\underline{k},\omega_n)\; G(-\underline{k},-\omega_n)\; \Sigma_{02}(\underline{k}) . \quad (5.4)$$

Resummation of the perturbation expansion for

$$G(\underline{k},\omega_n) = \frac{\omega_n + (k^2/2m) - \mu + \Sigma_{11}(\underline{k})}{\omega_n^2 - \left[(k^2/2m) - \mu + \Sigma_{11}(\underline{k})\right]^2 + \Sigma_{02}^2(\underline{k})} , \quad (5.5)$$

It is convenient to introduce the notation

$$\tilde{\epsilon}_{\underline{k}} = \frac{k^2}{2m} - \mu + \Sigma_{11}(\underline{k}) ,$$

$$\Delta_{\underline{k}} = \Sigma_{02}(\underline{k}) ,$$

$$\epsilon_{\underline{k}}^2 = \tilde{\epsilon}_{\underline{k}}^2 - \Delta_{\underline{k}}^2 . \quad (5.6)$$

Equation (5.5) therefore becomes

$$G(\underline{k},\omega_n) = \frac{1}{2}\left\{\left(1 + \frac{\tilde{\epsilon}_{\underline{k}}}{\epsilon_{\underline{k}}}\right)\frac{1}{\omega_n - \epsilon_{\underline{k}}} + \left(1 - \frac{\tilde{\epsilon}_{\underline{k}}}{\epsilon_{\underline{k}}}\right)\frac{1}{\omega_n + \epsilon_{\underline{k}}}\right\} . \quad (5.7)$$

To evaluate sums over frequencies we use the relation

$$-\frac{1}{\beta}\sum_{\omega_n} \frac{e^{\omega_n \tau}}{\omega_n - \epsilon_{\underline{k}}} = \frac{1}{2\pi i}\int_C \frac{dz}{e^{\beta z} - 1}\,\frac{1}{z - \epsilon_{\underline{k}}} = \frac{1}{e^{\beta\epsilon_{\underline{k}}} - 1} \quad (5.8)$$

since ω_n is an even multiple of $\pi i/\beta$. The contour C encircles the poles of $1/[e^{\beta z} - 1]$ in a clockwise direction, and may be deformed to encircle the pole at $z = \epsilon_{\underline{k}}$.

Using equations (5.7) and (5.8), we find

$$-\frac{1}{\beta}\sum_{\omega_n} G(\underline{k},\omega_n)\, e^{\omega_n \tau} = \frac{1}{2}\left(\frac{\tilde{\epsilon}_{\underline{k}}}{\epsilon_{\underline{k}}}\coth\tfrac{1}{2}\beta\epsilon_{\underline{k}} - 1\right). \tag{5.9}$$

Equations (5.3) and (5.6), therefore, give

$$\Sigma_{11}(\underline{p}) = \frac{N}{V}v(0) + \frac{1}{2V}\sum_{\underline{k}} v(\underline{p}-\underline{k})\left(\frac{\tilde{\epsilon}_{\underline{k}}}{\epsilon_{\underline{k}}}\coth\tfrac{1}{2}\beta\epsilon_{\underline{k}} - 1\right), \tag{5.10}$$

where $N = \sum_{\underline{k}} n_{\underline{k}}$.

Now we return to the equation (5.4) for Σ_{02}. Since

$$G_n^{-1}(\underline{p},\omega_n) = G_o^{-1}(\underline{p},\omega_n) - \Sigma_{11}(\underline{p},\omega_n) ,$$

we find

$$\frac{1}{\beta}\sum_{\omega_n} G_n(\underline{k},\omega_n)\, G(-\underline{k},-\omega_n) = -\frac{1}{\beta}\sum_{\omega_n}\frac{1}{\omega_n^2 - \epsilon_{\underline{k}}^2} = -\frac{1}{2\epsilon_{\underline{k}}}\coth\tfrac{1}{2}\beta\epsilon_{\underline{k}} . \tag{5.11}$$

Substituting equation (5.9) into equation (5.3) and equation (5.11) into equation (5.4), and using the relations (5.6), we see that

$$\tilde{\epsilon}_{\underline{k}} = \frac{k^2}{2m} - \mu + \frac{N}{V}v(0) + \frac{1}{2V}\sum_{\underline{p}} v(\underline{k}-\underline{p})\left(\frac{\tilde{\epsilon}_{\underline{p}}}{\epsilon_{\underline{p}}}\coth\tfrac{1}{2}\beta\epsilon_{\underline{p}} - 1\right),$$

$$\Delta_{\underline{k}} = -\frac{1}{2V}\sum_{\underline{p}} v(\underline{k}-\underline{p})\,\frac{\Delta_{\underline{p}}}{\epsilon_{\underline{p}}}\coth\tfrac{1}{2}\beta\epsilon_{\underline{p}} , \tag{5.12}$$

which are the same as the equations (2.14) obtained by decoupling Green functions, or from Luban's (1962) work.

There is one slight difference between these results and the earlier ones; the summations in equation (5.12) are not restricted. However, the changes introduced by this difference are only of order $1/V$ and may be neglected.

3.6 Luban's Solution of the Pair Hamiltonian Model

Earlier we remarked that difficulties are experienced when attempting to apply perturbation theory to the condensed Bose gas, and we now point out an error in Luban's work. Luban calculates the ratio $\mathrm{Tr}\exp(-\beta\hat{H}_p)/\mathrm{Tr}\exp(-\beta\hat{H}^o)$ by thermodynamic perturbation theory, and he wishes to show that each term in the perturbation expansion is independent of the volume. To do this, Luban needs to establish that $\langle(\hat{a}_o^+\hat{a}_o - n_o)^2\rangle^o$ is of order V as $V\to\infty$. Unfortunately, this result is incorrect due to the limit $V\to\infty$ being evaluated too soon. Luban argues that the free energy Ω is proportional to the volume, and, therefore, the derivatives of Ω with respect to μ must also share this property. Using this result he shows that the required result mentioned above is true. However, the free energy of a condensed Bose gas contains terms which are not analytic functions of V as $V \to \infty$; the value of these terms tends to zero as $V \to \infty$, but their derivatives with respect to the volume do not do so. As an example we quote the ideal Bose gas: the free energy due to the particles in the zero-momentum state vanishes as $V \to \infty$, whereas the derivatives with respect to the chemical potential, which is of order $1/V$, depend on a positive power of the volume.

In the ideal Bose gas $\langle(\hat{a}_0^+\hat{a}_0 - n_0)^2\rangle$ is of order V^2 as $V \to \infty$, and a similar result holds for the Bose gas described by $\hat{H}^o$. The fluctuation is easily evaluated by differentiating n_0 with respect to $\tilde{\varepsilon}_0$, if we use the relation (2.12) for n_0. Differentiating Ω with respect to μ after taking the limit $V \to \infty$ removes the dynamics of the zero-momentum state. For this problem the operations $\partial/\partial\mu$ and $\lim_{V\to\infty}$ do not commute.

As a result of these observations we see that Luban's solution has not been shown to be exact. In the next section we present an exact solution based on a method very close in spirit to Luban's original one.

3.7 An Exact Solution of the Pair Hamiltonian Model

To solve the problem exactly, we use a method closely related to Luban's own: we construct an approximating Hamiltonian and then show that its thermodynamic properties are the same as those of $\hat{H}_p$.

We first of all perform a canonical transformation on the zero-momentum operators:

$$\hat{a}_o \to \alpha V^{\frac{1}{2}} + \hat{a}_o \quad ,$$

$$\hat{a}_o^+ \to \alpha V^{\frac{1}{2}} + \hat{a}_o^+ \quad , \qquad (7.1)$$

where α is a real c-number. After we have performed the transformation (7.1), the pair Hamiltonian has the form

$$\hat{H}_p = \sum\left[T_{\underline{k}} + \alpha^2\left(v(0) + v(\underline{k})\right)\right]\hat{a}^+_{\underline{k}}\hat{a}_{\underline{k}} + \frac{v(0)}{V}\sum \hat{a}^+_{\underline{p}}\hat{a}^+_{\underline{k}}\hat{a}_{\underline{k}}\hat{a}_{\underline{p}}$$

$$+ \frac{1}{2V}\sum{}'' v(\underline{p}-\underline{k})\hat{a}^+_{\underline{p}}\hat{a}^+_{\underline{k}}\hat{a}_{\underline{k}}\hat{a}_{\underline{p}} + \frac{1}{2V}\sum{}' v(\underline{p}-\underline{k})\,\hat{a}^+_{\underline{p}}\hat{a}^+_{-\underline{p}}\hat{a}_{\underline{k}}\hat{a}_{-\underline{k}}$$

$$+ \left[T_0 + \alpha^2 v(0)\right]\alpha V^{\frac{1}{2}}(\hat{a}_0 + \hat{a}^+_0) + \frac{\alpha v(0)}{V^{\frac{1}{2}}}\sum(\hat{a}^+_{\underline{k}}\hat{a}_{\underline{k}}\hat{a}_0 + \hat{a}^+_0\hat{a}^+_{\underline{k}}\hat{a}_{\underline{k}}$$

$$+ \frac{\alpha}{V^{\frac{1}{2}}}\sum{}' v(\underline{k})(\hat{a}^+_{\underline{k}}\hat{a}_{\underline{k}}\hat{a}_0 + \hat{a}^+_0\hat{a}^+_{\underline{k}}\hat{a}_{\underline{k}}) + \frac{\alpha}{V^{\frac{1}{2}}}\sum{}' v(\underline{k})\,(\hat{a}^+_0\hat{a}_{\underline{k}}\hat{a}_{-\underline{k}} + \hat{a}^+_{\underline{k}}\hat{a}^+_{-\underline{k}}\hat{a}_0)$$

$$+ \frac{1}{2}\alpha^2\sum v(\underline{k})(\hat{a}^+_{\underline{k}}\hat{a}^+_{-\underline{k}} + \hat{a}_{-\underline{k}}\hat{a}_{\underline{k}}) + T_0\alpha^2 V + \tfrac{1}{2}\alpha^4 V v(0). \tag{7.2}$$

The transformation (7.1) is the one performed by Popov and Faddeev (1964) in their work on the Bose gas.

We introduce the notation

$$\left.\begin{aligned} n_{\underline{k}} &= \langle\hat{a}^+_{\underline{k}}\hat{a}_{\underline{k}}\rangle\ , \\ A_{\underline{k}} &= \langle\hat{a}^+_{\underline{k}}\hat{a}^+_{-\underline{k}}\rangle = \langle\hat{a}_{-\underline{k}}\hat{a}_{\underline{k}}\rangle\ , \\ \hat{B}_{\underline{k}} &= \hat{a}^+_{\underline{k}}\hat{a}^+_{-\underline{k}} - n_{\underline{k}}\ , \\ \hat{C}^+_{\underline{k}} &= \hat{a}^+_{\underline{k}}\hat{a}^+_{-\underline{k}} - A_{\underline{k}}\ , \\ \hat{C}_{\underline{k}} &= \hat{a}_{-\underline{k}}\hat{a}_{\underline{k}} - A_{\underline{k}}\ , \end{aligned}\right\} \tag{7.3}$$

where thermal averages are evaluated using the approximating Hamiltonian H'_{op} to be defined by equation (7.5). By using the definitions equation (7.3), we may write the pair Hamiltonian equation (7.2) in the form

$$\hat{H}_p = H'_{op} + \frac{v(0)}{2V}\sum\hat{B}_{\underline{p}}\hat{B}_{\underline{k}} + \frac{1}{2V}\sum{}'' v(\underline{p}-\underline{k})\hat{B}_{\underline{p}}\hat{B}_{\underline{k}}$$

$$+ \frac{1}{2V}\sum{}' v(\underline{p}-\underline{k})\,\hat{C}^+_{\underline{p}}\hat{C}_{\underline{k}} + \frac{\alpha v(0)}{V^{\frac{1}{2}}}\sum(\hat{B}_{\underline{k}}\hat{a}_0 + \hat{a}^+_0\hat{B}_{\underline{k}}) \tag{7.4}$$

$$+ \frac{\alpha}{V^{\frac{1}{2}}}\sum{}' v(\underline{k})\hat{B}_{\underline{k}}(\hat{a}_0 + \hat{a}^+_0) + \frac{\alpha}{V^{\frac{1}{2}}}\sum{}' v(\underline{k})(\hat{C}^+_{\underline{k}}\hat{a}_0 + \hat{a}^+_0\hat{C}_{\underline{k}})\ ,$$

where $(N' \equiv N - \alpha^2 V)$

$$H'_{op} = \sum \left\{ T_{\underline{k}} + \frac{N}{V} v(0) + \frac{1}{V} \sum{}'' v(\underline{k}-\underline{p})\, n_{\underline{p}} + \alpha^2 v(\underline{k}) \right\} \hat{a}^{+}_{\underline{k}} \hat{a}_{\underline{k}}$$

$$+ \sum \left\{ \frac{\alpha^2}{2} v(\underline{k}) + \frac{1}{2V} \sum{}' v(\underline{k}-\underline{p})\, A_{\underline{p}} \right\} (\hat{a}^{+}_{\underline{k}} \hat{a}^{+}_{-\underline{k}} + \hat{a}_{-\underline{k}} \hat{a}_{\underline{k}})$$

$$+ \alpha V^{\frac{1}{2}} \left\{ T_o + \frac{N}{V} v(0) + \frac{1}{V} \sum{}' v(\underline{k})\, n_{\underline{k}} + \frac{1}{V} \sum v(\underline{k})\, A_{\underline{k}} \right\} (\hat{a}_o + \hat{a}^{+}_o)$$

$$- \frac{N'^2}{2V} v(0) - \frac{1}{2V} \sum{}'' v(\underline{p}-\underline{k})\, n_{\underline{p}} n_{\underline{k}} - \frac{1}{2V} \sum{}' v(\underline{p}-\underline{k})\, A_{\underline{p}} A_{\underline{k}}$$

$$+ T_o \alpha^2 V + \frac{1}{2} \alpha^4 V v(0) . \tag{7.5}$$

Let us choose α such that $\langle \hat{a}_o \rangle = 0$. This condition is clearly satisfied if the coefficient of $(\hat{a}_o + \hat{a}^{+}_o)$ in H'_{op} is zero, since the other part of the Hamiltonian couples states whose numbers of particles differ by zero or two. Hence

$$T_o + \frac{N}{V} v(0) + \frac{1}{V} \sum{}' v(\underline{k})\, n_{\underline{k}} + \frac{1}{V} \sum{}' v(\underline{k})\, A_{\underline{k}} = 0 , \tag{7.6}$$

where we have rejected the solution $\alpha = 0$, since this corresponds to the uncondensed region. Equation (7.6) is equivalent to Luban's equation (39) expressing one of the consequences of B.E.-condensation; in our notation this reads

$$\tilde{\epsilon}_o + \Delta_o = 0 . \tag{7.7}$$

The Hamiltonian H'_{op} may now be diagonalized by a Bogolyubov transformation, since it is the same as the pair Hamiltonian. We find

$$n_{\underline{k}} = \frac{1}{2} \left[\frac{\tilde{\epsilon}'_{\underline{k}}}{\epsilon'_{\underline{k}}} \coth \tfrac{1}{2} \beta \epsilon'_{\underline{k}} - 1 \right] , \tag{7.8}$$

$$A_{\underline{k}} = -\frac{1}{2} \frac{\Delta'_{\underline{k}}}{\epsilon'_{\underline{k}}} \coth \tfrac{1}{2} \beta \epsilon'_{\underline{k}} ,$$

where

$$\tilde{\epsilon}'_{\underline{k}} = T_{\underline{k}} + \frac{N}{V} v(0) + \frac{1}{V} \sum{}' v(\underline{k}-\underline{p})\, n_{\underline{p}} + \alpha^2 v(\underline{k}) ,$$

$$\Delta'_{\underline{k}} = \frac{1}{V} \sum v(\underline{k}-\underline{p})\, A_{\underline{p}} + \alpha^2 v(\underline{k}) \tag{7.9}$$

and

$$\epsilon'^2_{\underline{k}} = \tilde{\epsilon}'^2_{\underline{k}} - \Delta'^2_{\underline{k}} .$$

Equations (7.8) and (7.9) may be used to give coupled integral equations for $\epsilon'_{\underline{k}}$ and $\Delta'_{\underline{k}}$:

$$\tilde{\epsilon}'_{\underline{k}} = T_{\underline{k}} + \frac{N}{V} v(0) + \frac{1}{2V} \sum{}' v(\underline{k}-\underline{p}) \left(\frac{\tilde{\epsilon}'_{\underline{p}}}{\epsilon'_{\underline{p}}} \coth \tfrac{1}{2} \beta \epsilon'_{\underline{p}} - 1 \right) + \alpha^2 v(\underline{k}) \; .$$

$$\Delta'_{\underline{k}} = -\frac{1}{2V} \sum v(\underline{k}-\underline{p}) \frac{\Delta'_{\underline{p}}}{\epsilon'_{\underline{p}}} \coth \tfrac{1}{2} \beta \epsilon_{\underline{p}} + \alpha^2 v(\underline{k}) \; . \tag{7.10}$$

The other two equations to be satisfied are equation (7.6) and the condition

$$N = \alpha^2 V + \frac{1}{2} \sum \left[\frac{\tilde{\epsilon}'_{\underline{p}}}{\epsilon'_{\underline{p}}} \coth \tfrac{1}{2} \beta \epsilon'_{\underline{p}} - 1 \right] \; , \tag{7.11}$$

which determines the chemical potential μ.

Although $\tilde{\epsilon}_0 + \Delta_0 = 0$, we see that $\tilde{\epsilon}'_0 + \Delta'_0 \neq 0$. Consequently $\epsilon'_0 \neq 0$, and there will not be large fluctuations in the zero-momentum state. We also see that $\epsilon'_{\underline{k}}$ is continuous at $\underline{k} = 0$, whereas $\epsilon_{\underline{k}}$ is discontinuous.

To conclude our proof, we must show that the interaction does not produce any change in the free energy of the system in the infinite volume limit. The terms in $\hat{H}_p - H'_{op}$, which are given by equation (7.4), are of two types: $(\alpha/V^{\frac{1}{2}})\, v D\, (\hat{a}_0 + \hat{a}_0^+)$ and $v \hat{D} \hat{D}/V$ where v is some matrix element of the interaction potential and $\hat{D}$ stands for $\hat{B}_{\underline{k}}$, $\hat{C}^+_{\underline{k}}$, or $\hat{C}_{\underline{k}}$. A typical term in the expansion of $\mathrm{Tr}\, \exp(-\beta \hat{H}_p)/\mathrm{Tr}\, \exp(-\beta H'_{op})$ will have the form

$$\frac{1}{V^{n+m/2}} \sum v^{n+m} \langle (\hat{D}^2)^n (\hat{D}\, \hat{a}_0)^m \rangle \; .$$

The total number of momentum summations is $2n + m$, but not all of these are independent. The thermal average may be factorized; if there is only a single $\hat{D}$ factor of some momentum, the average will vanish, since $\langle \hat{D} \rangle = 0$. The largest number of pairs of $\hat{D}$ factors that can be formed is $n + \frac{m}{2}$ for m even and $n + \frac{m-1}{2}$ for m odd. Thus the total number of independent summations over momentum-space is at most $n + \frac{m}{2}$. The term in the perturbation theory expansion is therefore volume-independent since no correlation functions depend on a positive power of the volume.

We conclude that all terms in the perturbation expansion of $\mathrm{Tr}\, \exp(-\beta \hat{H}_p)/\mathrm{Tr}\, \exp(-\beta H_{op})$ are volume-independent, and hence the difference between the free energies of the two systems is independent of the volume. Consequently, our solution of the pair Hamiltonian problem is asymptotically exact in the infinite volume limit.

References

Abrikosov, A.A., Gor'kov, L.P. and Dzyaloshinskii, I.E., Quantum Field Theoretical Methods in Statistical Mechanics, 2nd edition, Pergamon, Oxford, 1965.

Bogolyubov, N.N., Physica, **26**, S 1 (1960).

Bogolyubov, N.N., Zubarev, D.N. and Tserkovnikov, Yu., JETP, **39**, 120 (Soviet Physics - JETP, **12**, 88 (1961)).

Girardeau, M., J. Math. Phys., **3**, 131 (1962).

Girardeau, M. and Arnowitt, R., Phys. Rev., **113**, 755 (1959).

ter Haar, D., In Fluctuation, Relaxation and Resonance in Magnetic Systems, Oliver and Boyd, Edinburgh, p.119, 1962.

Hugenholtz, N.M. and Pines, D., Phys. Rev, **116**, 489 (1959).

Luban, M., Phys. Rev., **128**, 965 (1962).

Pines, D., The Many-Body Problem, Benjamin, New York, p.98, 1961.

Popov, V.N. and Faddeev, L.D., JETP, **47**, 1315 (Soviet Physics - JETP, **20**, 890 (1965)).

Valatin, J.G. and Butler, D., Il Nuovo Cimento, **10**, 37 (1958).

Wentzel, G., Phys. Rev., **120**, 1572 (1960).

Zubarev, D.N., Usp. Fiz. Nauk., **71**, 71 (Soviet Physics - Uspekhi, **3**, 320 (1960))

Zubarev, D.N. and Tserkovnikov, Yu.A., Dokl. Akad. Nauk. SSSR, **120**, 991 (Soviet Physics - Doklady, **3**, 603 (1958)).

Appendix – The Bogolyubov Hamiltonian and Transformation

Consider the following Hamiltonian of a system of interacting bosons (compare equation (1.50) of chapter 1

$$H_{op} = \sum \epsilon_{\underline{k}} \hat{a}^{+}_{\underline{k}} \hat{a}_{\underline{k}} + \frac{1}{2V} \sum_{\underline{k},\underline{\ell},\underline{q}} v(\underline{q})\, \hat{a}_{\underline{k}+\underline{q}} \hat{a}_{\underline{\ell}+\underline{q}} \hat{a}_{\underline{\ell}} \hat{a}_{\underline{k}} , \tag{1}$$

where ($\hbar = 1$)

$$\epsilon_{\underline{k}} = \frac{k^2}{2m} . \tag{2}$$

and where that $\hat{a}_{\underline{k}}$ and $\hat{a}^{+}_{\underline{k}}$ are the boson annihilation and creation operators satisfying the usual commutation relations. The problem posed by the Hamiltonian (1) is insoluble, but Bogolyubov has shown how a simplified model may be solved. At low temperatures one expects that a finite fraction of the bosons will be in the zero-momentum state, that is, $\langle \hat{a}^{+}_{0} \hat{a}_{0} \rangle \equiv N_0$ is of the same order as the total number of particles, even though $\langle \hat{a}^{+}_{\underline{k}} \hat{a}_{\underline{k}} \rangle$ for $k \neq 0$ will be of the order of unity, that is, smaller than N_0 by a factor of order N. This means that to a good approximation we can treat $\hat{a}^{+}_{0}$ and $\hat{a}_{0}$ to be c-numbers, both equal to $\sqrt{N_0}$. Bogolyubov, therefore, retains in (1) only those terms in which at most two of the momenta differ from zero. This means that he replaces (1) by the Hamiltonian

$$H'_{op} = \sum \epsilon_{\underline{k}} \hat{a}^+_{\underline{k}} \hat{a}_{\underline{k}} + \frac{N_o^2}{2V} v(0) + \frac{N_o}{2V} \sum_{q(\neq 0)} v(\underline{q}) (\hat{a}^+_{\underline{q}} \hat{a}^+_{-\underline{q}} + \hat{a}_{\underline{q}} \hat{a}_{-\underline{q}})$$
$$+ \frac{N_o}{V} v(0) \sum_{\underline{q}(\neq 0)} \hat{a}^+_{\underline{q}} \hat{a}_{\underline{q}} + \frac{N_o}{V} \sum_{\underline{q}(\neq 0)} v(\underline{q}) \hat{a}^+_{\underline{q}} \hat{a}_{\underline{q}} . \quad (3)$$

Again neglecting terms containing more than two non-zero momenta, we can write (3) in the form

$$H'_{op} = \sum_{\underline{q}} \epsilon_{\underline{q}} \hat{a}^+_{\underline{q}} \hat{a}_{\underline{q}} + \frac{N^2}{2V} v(0) + \frac{N}{2V} \sum_{q(\neq 0)} v(\underline{q}) \left[2 \hat{a}^+_{\underline{q}} \hat{a}_{\underline{q}} + \hat{a}_{\underline{q}} \hat{a}^+_{\underline{q}} + \hat{a}_{\underline{q}} \hat{a}_{-\underline{q}} \right] . \quad (4)$$

In order to bring this Hamiltonian in an easily manageable form, Bogolyubov introduced a canonical transformation to new boson operators, $\hat{b}_{\underline{q}}$ and $\hat{b}_{\underline{q}}$,

$$\hat{a}_{\underline{q}} = \alpha_{\underline{q}} \hat{b}_{\underline{q}} + \beta_{\underline{q}} \hat{b}^+_{-\underline{q}} \quad , \quad \hat{a}^+_{\underline{q}} = \alpha_{\underline{q}} \hat{b}^+_{\underline{q}} + \beta_{\underline{q}} \hat{b}_{-\underline{q}} \quad , \quad (5)$$

with real $\alpha_{\underline{q}}$ and $\beta_{\underline{q}}$. The requirement that the $\hat{b}_{\underline{q}}$ and $\hat{b}^+_{\underline{q}}$ satisfy the same commutation rules as the $\hat{a}_{\underline{q}}$ and $\hat{a}^+_{\underline{q}}$ gives us one relation between $\alpha_{\underline{q}}$ and $\beta_{\underline{q}}$ and the requirement that the coefficient of the terms with $\hat{b}^+_{\underline{q}} \hat{b}^+_{-\underline{q}}$ and $\hat{b}_{\underline{q}} \hat{b}_{-\underline{q}}$ must vanish gives us a second relation. Having thus determined the $\alpha_{\underline{q}}$ and $\beta_{\underline{q}}$ we can transform the Hamiltonian, and the result is

$$H'_{op} = \sum_{\underline{q}} E_{\underline{q}} \hat{b}^+_{\underline{q}} \hat{b}_{\underline{q}} + \text{constant} , \quad (6)$$

which is the Hamiltonian of a system of independent quasi-particles. The quasi-particle energy $E_{\underline{q}}$ is given by the equation

$$E_{\underline{q}} = \left[\frac{q^2 v(q) N}{mV} + \frac{q^4}{4m^2} \right]^{\frac{1}{2}} , \quad (7)$$

which for small q leads to

$$E_{\underline{q}} \approx sq \quad , \quad s = \left(\frac{v(0)N}{mV} \right)^{\frac{1}{2}} , \quad (8)$$

that is, the phonon spectrum.

CHAPTER 4
Fluctuations in a Perfect Bose Gas

4.1 The Use of Gentile Statistics

The standard expression for the fluctuation of the occupation numbers in Bose statistics,

$$\overline{\left(\frac{\Delta n}{\bar{n}}\right)^2} = 1 + \frac{1}{\bar{n}} , \qquad (1.1)$$

meets with difficulties when applied to the ground state at very low temperatures. As an asymptotic formula for this special case, we propose instead

$$\left(\frac{\Delta n_0}{\bar{n}_0}\right)^2 \rightarrow \left[\frac{N}{\bar{n}_0} - 1\right]^2 , \qquad (1.2)$$

as $T \to 0$. In contrast to formula (1.1), this fulfils the fundamental requirement that the fluctuations in the population of the lowest energy level must go to zero with the temperature — in Bose statistics, no less than in Fermi statistics.

To be sure, formula (1.1) which implies relative fluctuations of about 100% is 'exact' in as much as it follows without approximation from the grand canonical ensemble. This ensemble is tacitly implied in most current formulations of quantum statistics; and as a rule, there are no observable differences in the results obtained from the various ensembles. This is certainly true for Fermi statistics, and for Bose statistics the only exception is when we consider the ground state at very low temperatures.

In this particular case, however, it must be remembered that, strictly speaking, the grand canonical ensemble describes open systems only — although incidentally it almost always gives correct formulas even for closed systems. The large fluctuations in n_0 indicated by equation (1.1) thus refer to systems which can freely exchange particles with an infinite reservoir or with another phase. But this picture is not at all appropriate to experimental situations, where the system is a macroscopically conserved amount of matter: in Bose statistics, n_0 may, at sufficiently low temperature, be a macroscopic quantity and its relative fluctuation must certainly vanish lest the theory predict very startling phenomena.

For a proper treatment of the present problem, it is thus necessary to take conservation of the total number of particles N into account. This will be shown to require, for the ground state at low temperatures, equation (1.2) instead of equation (1.1).

Before discussing specific examples, we shall write down a description of the canonical ensemble which is suitable for the present purpose: Let the molecular energy levels in a closed system of N non-interacting particles be ϵ_s, $s = 0, 1, 2, \ldots$, but we shall now consider the total number of particles

$$\sum_s n_s = N \tag{1.3}$$

to be fixed in the strong sense — not merely in canonical average. Accordingly, the sum over states must be restricted:

$$Z = \sum_{n_s}{}' \prod_s \exp(-\beta n_s \epsilon_s) . \tag{1.4}$$

Here, Σ' denotes a multiple sum over only those sets of occupation numbers $\{n_s\}$ that satisfy the condition (1.3).

Such a condition is conveniently imposed upon the summation in equation (1.4) by multiplying each term of the unrestricted sum by a discontinuous factor:

$$\frac{1}{2\pi i}\int_{-\pi i}^{\pi i} dt \exp\left\{-\left(N - \sum_s n_s\right)\right\} t = \begin{cases} 1, & \text{when (1.3) is fulfilled,} \\ 0, & \text{otherwise .} \end{cases}$$

In this way, the sum over states can be written as an integral

$$Z = \frac{1}{2\pi i}\int dt\; e^{-Nt} \prod_s \sum_{n_s} \left\{\exp\left[n_s(t - \beta\epsilon_s)\right]\right\}, \tag{1.5}$$

which can be evaluated by the Darwin-Fowler method.

Usually, the sums over the n_s are understood to be from 0 to ∞,

$$\sum_{n_s=0}^{\infty} = 1/\left(1 - e^{t-\beta\epsilon_s}\right) , \tag{1.6}$$

but this is not the only possibility: we may truncate them at any number $M \geq N$, inserting, instead of the sum (1.6),

$$\sum_{n_s=0}^{M} = \left(1 - \exp\left[(M_s + 1)(t - \beta\epsilon_s)\right]\right)/\left(1 - \exp(t - \beta\epsilon_s)\right) \tag{1.7}$$

for the sth factor in the integrand of equation (1.5). So far, the replacements of the infinite series [equation (1.6)] by finite sums [equation (1.7)] are quite arbitrary. They cannot alter the exact value of the integral [equation (1.5)]. In most cases, however, only asymptotic approximations for large N are available, and in order to obtain them it will sometimes be a considerable advantage to take the factors of the infinite product in the form (1.7), which is an entire function of t, instead of the form (1.6), which is not. This will turn out to 'temper' the analytical properties of the integrand, and thus to make it more amenable to a saddle-point approximation. We shall here choose all $M_s = N$, that is, pick the smoothest form possible (salva veritate) for the integrand in equation (1.5).

Incidentally, this choice $(M = N)$ leads to the particular form of distribution law for the occupation numbers which was proposed by Gentile on somewhat different grounds. Although Gentile's procedure, equation (1.7), cannot claim to be more than a computational device, we shall see that it is useful for boson systems at very low temperatures. Actually, it extends the applicability of the method of steepest descent to a range of states where in the usual formalism this method was very problematic. For the purposes of our fluctuation problem, we shall, therefore, adopt his version of the sum over states,

$$Z = \frac{1}{2\pi i} \int dt \ e^{-Nt} \prod_s \left\{\left(1 - e^{-Dz_s}\right)/\left(1 - e^{-z_s}\right)\right\} \tag{1.8}$$

where $D = N + 1$, $z = \beta \epsilon_s - t$. By the definition (1.4) of our ensemble, the average occupation numbers and their fluctuations are given by

$$\bar{n} = - \frac{\partial \ln Z}{\partial \beta \epsilon} \ , \quad \overline{\Delta n^2} = \frac{\partial^2 \ln Z}{\partial (\beta\epsilon)^2} , \tag{1.9}$$

respectively.

Using the saddle-point approximation, we get from equation (1.9)

$$\ln Z = - Nt + \sum_s \chi(\beta\epsilon_s - t) \ , \tag{1.10}$$

where

$$\chi(z) = - \ln(1 - e^{-z}) + \ln(1 - e^{-Dz}) \ ,$$

and terms of the order $\ln N$ have been neglected. The principal saddle-point t is the real root of the equation

$$N = \sum_s \dot{\chi}_s \ , \quad \dot{\chi}_s = \frac{\partial \chi}{\partial t} = - \chi' \tag{1.11}$$

While χ is a universal function, the parameter t depends specifically upon the system through the energy spectrum ϵ_s. Inserting the approximation (1.10) in the general equations (1.9), one thus obtains

$$\bar{n}_s = \dot{\chi}_s , \quad \overline{\Delta n_s^2} = \ddot{\chi}_s \left[1 - \frac{\partial t}{\partial \beta \epsilon_s} \right] . \tag{1.12}$$

These formulas are of a well-known type, except that Gentile's expression

$$\dot{\chi}(z) = \frac{1}{e^z - 1} - \frac{D}{e^{Dz} - 1} , \quad z = \beta\epsilon - t , \tag{1.13}$$

differs from the familiar distribution law of Bose statistics,

$$f(z) = \frac{1}{e^z - 1} , \tag{1.14}$$

by the extra term $-Df(Dz)$. This term, however, makes the function $\bar{n} = \dot{\chi}(z)$ always finite, positive, and monotonically decreasing:

$$\dot{\chi}(-\infty) = N , \quad \dot{\chi}(0) = N(0) , \quad \dot{\chi}(\infty) = 0 , \tag{1.15}$$

with a steep drop at $z = 0$, as can be shown.

By the usual thermodynamic identification, t is related to the chemical potential $(t = \beta\mu)$. At ordinary temperatures, the chemical potential is negative whether we define it by the conventional means or by Gentile's formula. Then, $z = \beta\epsilon - t$ is positive for all quantum states and the term $-Df(Dz)$ is completely negligible, as of course it must be in this case. In the usual formalism, which employs the distribution law (1.14), z must in fact always be positive, that is, t can never exceed $\beta\epsilon_0$.

Gentile's function $\bar{n} = \dot{\chi}(z)$, on the other hand, is always finite and z may well assume negative values, as indeed it does for the lowest quantum state at sufficiently low temperatures. Thus, t may exceed $\beta\epsilon_0$ although it must always remain below $\beta\epsilon_1$ [$t > \beta\epsilon_1 > \beta\epsilon_0$ violates equation (1.11) (cf. equations (1.15))]. The thermodynamic states in which we are here primarily interested are just those belonging to positive values of t: $0 \le t < \beta\epsilon_1$. When $t = \beta\epsilon_0$, we have $\bar{n}_0 = \dot{\chi}(0) = N/2$: one half of the particles are in the ground state. At absolute zero, $\beta = \infty$, we shall have $t = +\infty$, and $\overline{n_0} = N$, as must be the case.

Our main concern, however, is the fluctuations. These depend upon the function $\ddot{\chi}_0$ entering in the formula

$$\overline{\Delta n_0^2} = \ddot{\chi}_0 \left[1 - \frac{\partial t}{\partial \beta \epsilon_0} \right] \tag{1.16}$$

which, as we shall see, embodies both equations (1.1) and (1.2) — the former at ordinary temperatures and the latter in the limit $T \to 0$.

In order to make this plausible, we use the identity, which we shall not prove here,

$$\ddot{\chi}_0 = D\bar{n}_0 - \bar{n}_0^2 + (2n_0 - N)\ f(z_0)\ . \tag{1.17}$$

At ordinary temperatures, when the term $Df(Dz)$ in equation (1.13) can be neglected, we have clearly $f(z_0) = n_0$, and accordingly,

$$\ddot{\chi}_0 = (\bar{n}_0)^2 + \bar{n}_0\ ,$$

which leads to equation (1.1). At sufficiently low temperatures, however, z will be negative and we may have, by equations (1.12) and (1.13), $f(z_0) = \bar{n}_0 - N + \mathcal{O}(N)$, while $\bar{n}_0 = \mathcal{O}(N)$, that is

$$\ddot{\chi}_0 = (N - \bar{n}_0)^2 + \mathcal{O}(N)\ , \tag{1.18}$$

and this gives formula (1.2).

So far, we have disregarded the factor $1 - \partial t/\partial\beta\varepsilon$ in equation (1.12). By variation of the saddle-point equation (1.11) with respect to the spectrum, one finds

$$\left(\frac{\partial t}{\partial\beta\varepsilon_0}\right)_N = \frac{\ddot{\chi}_0}{\sum_s \ddot{\chi}_s} < 1\ . \tag{1.19}$$

Whatever the magnitude of $\partial t/\partial\beta\varepsilon$, this term can, therefore, only decrease the fluctuation. At ordinary temperatures, it is obviously negligible, and as we shall see, also at sufficiently low temperatures. However, as was pointed out by Hiis Hauge, the factor $1 - \partial t/\partial\beta\varepsilon_0$ may be extremely important: he showed in the case of the ideal Bose gas that this factor will effectively suppress the fluctuation to zero at all temperatures below the Einstein transition point although $\ddot{\chi}_0$ is of the order N^2 in most of this range. But in any case, $\partial t/\partial\beta\varepsilon$ will depend upon the spectrum and, as will be shown for the ideal gas, the fluctuation formula (1.2) subsists in the limit as $T \to 0$.

According to equation (1.11), the saddle-point t will be determined by

$$N = \sum_\varepsilon \left[\frac{1}{e^{\beta\varepsilon - t} - 1} - \frac{D}{e^{D(\beta\varepsilon - t)} - 1}\right]. \tag{1.20}$$

Now, for macroscopic volumes V, the energy levels of an ideal gas are so densely spaced that, even at the lowest attainable temperatures, they cannot be resolved in thermal experiments. One is, therefore, inclined to replace sums by integrals :

$$\sum_{\epsilon} \rightarrow \frac{\beta^{\frac{3}{2}}}{\Gamma(\frac{3}{2})} \frac{V}{\lambda^3} \int \epsilon^{\frac{1}{2}} d\epsilon \,, \tag{1.21}$$

where $\lambda = (2\pi\hbar^2 \beta/m)$ is the thermal de Broglie wavelength. In the usual formulation of Bose statistics, this causes a well-known difficulty because the function

$$I(t) = \frac{1}{\Gamma(\frac{3}{2})} \int \frac{x^{\frac{1}{2}} dx}{e^{x-t} - 1} \tag{1.22}$$

has an essential singularity at $t = 0$. In Gentile's formulation, however, this is exactly compensated by the second sum in the curly bracket and thus the integral

$$N = \frac{V}{\lambda^3} \frac{1}{\Gamma(\frac{3}{2})} \int_0^\infty x^{\frac{1}{2}} dx \, \dot{\chi}(x-t) \tag{1.23}$$

is regular along the entire real t axis. We may, therefore, with some confidence, use equation (1.23) to determine the saddle-point. At temperatures above the Einstein transition t will be negative. This is the domain of validity of equation (1.1), which does not interest us here. Upon cooling of the system, the saddle-point will (ceteris paribus) move to the right along the real axis, reaching $t = 0$ at the temperature

$$T_E = \left[\frac{N\lambda^3 T^{\frac{3}{2}}}{\zeta(\frac{3}{2}) V}\right]^{\frac{2}{3}} . \tag{1.24}$$

At temperatures $T < T_E$, the saddle-point appears on the positive real axis and its location can be shown to be roughly given by

$$t \sim \left[1 - \left(\frac{T}{T_E}\right)^{\frac{3}{2}}\right] \frac{\lambda^2}{V^{\frac{2}{3}}} \,, \tag{1.25}$$

which is of the order $N^{-\frac{2}{3}}$ for moderately low temperatures $T \leqslant T_E$. To obtain a more accurate value, one must solve the equation

$$N = \dot{\chi}(z) + \frac{V}{\lambda^3} \zeta(\tfrac{3}{2}) + \Theta(N) \tag{1.26}$$

for $t = \beta\epsilon_0 - z$. Since $\dot{\chi}_0 = \bar{n}_0$ and $V\zeta(\frac{3}{2})/\lambda^3 = (T/T_E)^{\frac{3}{2}}$, this is the same as London's equation

$$\bar{n}_0 = \left[1 - \left(\frac{T}{T_E}\right)^{\frac{3}{2}}\right] N \,, \tag{1.27}$$

which is thus true in a quotient asymptotic sense.

It is not difficult to map the whole range of temperature $(0, T_E)$ upon the range $(\infty, 0)$ of t, but such detail is not necessary for our discussion of the relative fluctuation:

$$\left(\frac{\Delta n_0}{\bar{n}_0}\right)^2 = \frac{\ddot{\chi}_0}{\bar{n}_0^2}\left[1 - \frac{\partial t}{\partial \beta \epsilon_0}\right] . \tag{1.28}$$

All we need is a survey of $\bar{n}_0$, $\ddot{\chi}_0$, and $\partial t/\partial \beta \epsilon_0$ in their dependence upon the state.

For the ideal gas, one can prove that

$$\frac{\partial t}{\partial \beta \epsilon_0} \sim \frac{\ddot{\chi}_0}{\ddot{\chi}_0 + NV^{\frac{2}{3}}\lambda^{-2}} . \tag{1.29}$$

At the transition point $T = T_E$ we have

$$\ddot{\chi}_0 \sim N^{\frac{4}{3}} , \quad \bar{n}_0 \sim N^{\frac{2}{3}} , \quad V/\lambda^3 \sim N . \tag{1.30}$$

The relative fluctuation will, therefore, be of order 1, or, more precisely, it is just given by the standard formula (1.1).

In the middle of the range, $T \leqslant T_E[\circledS(z) \sim N^{-1}]$, we have

$$\ddot{\chi}_0 \sim N^2 , \quad \bar{n}_0 \sim N , \quad V/\lambda^3 \sim N , \tag{1.31}$$

so that $\ddot{\chi}_0/(\bar{n}_0)^2$ is still of order 1. Here, however, $\partial t/\partial\beta\epsilon_0$ is indeed close to 1, as observed by Hiis Hauge. From the relations (1.29) and (1.31) we have

$$1 - \frac{\partial t}{\partial \beta \epsilon_0} \sim \frac{N^{\frac{5}{3}}}{N^2 + N^{\frac{5}{3}}} \sim N^{-\frac{1}{3}} ; \tag{1.32}$$

hence the relative fluctuation vanishes to this order. At still lower temperatures, z is negative and ultimately of larger order than N^{-1}. In this case, the order of $\ddot{\chi}_0$ is given by the equation

$$\ddot{\chi}_0 \sim (N - n_0)^2 \sim (V/\lambda^{\frac{3}{2}})^2 . \tag{1.33}$$

The magnitude of $\partial t/\partial\beta\epsilon_0$ then depends upon the ratio

$$\left(\frac{V}{\lambda^3}\right)^2 \Big/ N\left(\frac{V}{\lambda^3}\right)^{\frac{2}{3}} \sim N^{\frac{1}{3}}\left(\frac{T}{T_E}\right)^2 . \tag{1.34}$$

That is, for temperatures $T \leqslant T_E N^{-\frac{1}{6}}$, the factor $1 - \partial t/\partial\beta\epsilon_0$ is of the order 1 and the fluctuation approaches

$$\left(\frac{\Delta n}{\bar{n}_0}\right)^2 \sim \left[\frac{N}{\bar{n}_0} - 1\right]^2 , \tag{1.35}$$

in agreement with formula (1.2). These are certainly very low temperatures, but it is noteworthy that they vastly exceed the range of temperatures $T < T_E N^{-\frac{2}{3}}$ envisaged in Planck's formulation of the third law.

For details of the calculations, we refer to the paper by Fujiwara, ter Haar, and Wergeland.

4.2 Symmetry Breaking in the Ground State of a Bose Gas

In Chapter 3 we briefly mentioned the so-called quasi-averages. Let us briefly discuss this topic. From the Landau theory of second-order phase transitions we know the important rôle played by the breaking of symmetry. For instance, the Heisenberg ferromagnet described in Chapter 2 in zero field is described by a rotationally invariant Hamiltonian, but below the Curie temperature it is characterized by a non-vanishing magnetization in a certain direction: there is an ordered state characterized by the existence of a non-vanishing order parameter, demonstrating that somehow or other a system with a Hamiltonian possessing a certain symmetry can get into a state which does not exhibit that symmetry.

Bogolyubov discussed this kind of situation by introducing into the Hamiltonian an infinitesimal term which breaks the symmetry, then evaluating thermodynamic averages, and after that removing the symmetry breaking field.

In the case of a ferromagnet this procedure consists in the following steps: one adds a term to the Hamiltonian which describes the interaction with an infinitesimal magnetic field $\varepsilon\underline{B}$, one calculates the magnetization $\underline{M}$ in the presence of this field (let us denote the resultant magnetization by $\underline{M}_\varepsilon$) and then one lets ε tend to zero. One then finds that below the Curie temperature the limit of $\underline{M}_\varepsilon$ as $\varepsilon \to 0$ is finite:

$$\underline{M} = \lim_{\varepsilon\to 0} \underline{M}_\varepsilon = \lim_{\varepsilon\to 0}\lim_{V\to\infty} \langle \hat{\underline{M}}_\varepsilon \rangle \neq 0 , \tag{2.1}$$

where the $\lim_{V\to\infty}$ indicates that we are taking the thermodynamic limit, and where the pointed brackets indicate once again a grand-canonical average.

Of course, if we had not had introduced the symmetry-breaking field, we would have found

$$\lim_{V\to\infty} \langle \hat{\underline{M}} \rangle_{\varepsilon=0} = 0 . \tag{2.2}$$

Moreover, the direction of the symmetry-breaking field determines the direction of $\underline{M}$, and in agreement with the rotational symmetry of the Hamiltonian for the case where $\varepsilon = 0$ we find that if we average $\underline{M}$ over all directions of $\varepsilon\underline{B}$, the result is zero.

Let us now turn to the case of a Bose gas, and let us consider the wavefunction operator ψ_{op} given by the equation

$$\psi_{op} = V^{-\frac{1}{2}} \sum_{\underline{k}} \hat{a}_{\underline{k}} \, e^{i(\underline{k}.\underline{r})} \quad . \tag{2.3}$$

From the physical meaning of the $\hat{a}_{\underline{k}}$ it follows that the operator ψ_{op} creates a particle at the position $\underline{r}$. Let us consider the case of a perfect Bose gas, where the operator H'_{op} occurring in the grand-canonical averages has the form given by equation (2.1) of Chapter 2 :

$$H'_{op} = \sum_{\underline{k}} \left[\epsilon(\underline{k}) - \mu \right] \hat{a}^{+}_{\underline{k}} \hat{a}_{\underline{k}} \quad . \tag{2.4}$$

As this Hamiltonian conserves the number of particles, we find that

$$\lim_{V \to \infty} \langle \psi_{op} \rangle_{\epsilon = 0} = 0 \ , \tag{2.5}$$

where ϵ measured the strength of a symmetry-breaking field which we still have to introduce.

On the other hand, we know that at sufficiently low temperatures Bose-Einstein condensation will occur which means that somehow or other some quasi-average of ψ must be non-vanishing. Before introducing a symmetry-breaking field let us just remind ourselves that the symmetry we have here is a gauge symmetry. The operator $\hat{H}'$ is invariant under the gauge transformation

$$\hat{a}_{\underline{k}} \longrightarrow \hat{a}_{\underline{k}} \, e^{i\chi} \ , \quad \hat{a}^{+}_{\underline{k}} \longrightarrow \hat{a}^{+}_{\underline{k}} \, e^{-i\chi} \quad . \tag{2.6}$$

We shall, therefore, introduce into the Hamiltonian a term which is not invariant under the transformation (2.6), and write

$$H'_{op}(\epsilon) = \sum_{\underline{k}} \left[\epsilon(\underline{k}) - \mu \right] \hat{a}^{+}_{\underline{k}} \hat{a}_{\underline{k}} + \epsilon V^{\frac{1}{2}} \left[e^{i\varphi} \hat{a}_{o} + e^{-i\varphi} \hat{a}^{+}_{o} \right] \tag{2.7}$$

where ϵ and φ are real quantities.

We would now expect — and shall, indeed, find — that

$$\psi = \lim_{\substack{\epsilon \to 0 \\ \varphi \text{ fixed}}} \psi_{\epsilon,\varphi} = \lim_{\substack{\epsilon \to 0 \\ \varphi \text{ fixed}}} \langle \psi_{op} \rangle_{\epsilon,\varphi} \neq 0 \ , \tag{2.8}$$

and also that, if we take the average of ψ over all possible values of φ we would get zero.

Before looking at the consequences of introducing $H'_{op}(\epsilon)$ we shall briefly recapitulate the results which we obtain with H'_{op}, and we shall concentrate on temperatures below the condensation temperature.

The grand partition function of a perfect Bose gas is given by the equation

$$Z_{gr} = \mathrm{Tr}\, \exp\left[-\beta H'_{op}\right] = \prod_{\underline{k}} \frac{1}{1 - ze^{-\beta\varepsilon(\underline{k})}} , \tag{2.9}$$

where

$$z = e^{\nu} , \quad \nu = \beta\mu . \tag{2.10}$$

We are interested in density fluctuations, and we need the formulae for the particle density and its fluctuations. The particle density follows from the equation:

$$\begin{aligned} n &= \lim_{V\to\infty} \langle n_{op}\rangle_V = \lim_{V\to\infty} \frac{1}{V} \langle N_{op}\rangle \\ &= \lim_{V\to\infty} \frac{1}{V}\frac{\partial}{\partial\nu} \ln Z_{gr} \\ &= \lim_{V\to\infty} \left[\frac{1}{V}\frac{z}{1-z} + \frac{1}{v_0} g_{\frac{3}{2}}(z)\right] , \end{aligned} \tag{2.11}$$

where v_0 is given by equation 1(1.11) :

$$v_0 = \left(\frac{2\pi\beta\hbar^2}{m}\right)^{\frac{3}{4}} , \tag{2.12}$$

and

$$g_n(z) = \sum_{\ell=1}^{\infty} \frac{z^{\ell}}{\ell^{n}} . \tag{2.13}$$

As $T\to 0$, $z\to 1$ so that it is convenient to use the variable $1-z=x$ which will be a small variable. In fact, as $g_{\frac{3}{2}}(1) = \zeta(\frac{3}{2})$, one finds from equation (2.11) in the limit as $V\to\infty$

$$x = \frac{1}{V}\left[n - \frac{\zeta(\frac{3}{2})}{v_0}\right]^{-1} . \tag{2.14}$$

We also note that the first term in the square brackets in equation (2.11) is n_0, that is, the density of zero-momentum particles, and from equation (2.14) it follows that

$$n_0 = \lim_{V\to\infty} \frac{1}{Vx} = n - \frac{\zeta(\frac{3}{2})}{v_0} . \tag{2.15}$$

To find the density fluctuations we use the equation

$$\begin{aligned} \langle(n - \langle n\rangle)^2\rangle &= \lim_{V\to\infty} \frac{1}{V^2} \langle(N_{op} - \langle n\rangle V)^2\rangle \\ &= \lim_{V\to\infty} \frac{1}{V^2} \frac{\partial^2}{\partial\nu^2} \ln Z_{gr} \\ &= \lim_{V\to\infty} \left[\frac{1}{V^2} \frac{z}{(1-z)^2} + \frac{1}{Vv_0} g_{\frac{1}{2}}(z)\right] , \end{aligned} \tag{2.16}$$

where we have dropped a term of order V^{-2} in the square brackets.

From equation (2.16) we find for $T < T_c$

$$\langle(n-\langle n\rangle)^2\rangle = \lim_{V\to\infty}\left[\frac{1}{V^2}V^2\left(n-\frac{\zeta(\frac{3}{2})}{v_0}\right)^2\right] = n_0^2 , \tag{2.17}$$

that is, we find macroscopic fluctuations in the density, in agreement with equation (1.1).

To see that these fluctuations come from the zero-momentum state, we note that, in fact, the first term within the square brackets in equation (2.17) equals the fluctuations in the zero-momentum state:

$$\lim_{V\to\infty}\frac{1}{V^2}\langle\{\hat{a}_0^+\hat{a}_0-\langle\hat{a}_0^+\hat{a}_0\rangle\}^2\rangle = \lim_{V\to\infty}\frac{1}{V^2}\frac{\partial}{\partial\nu}\langle N_0\rangle = \lim_{V\to\infty}\frac{1}{V^2}\frac{z}{(1-z)^2} . \tag{2.18}$$

Let us consider what happens when we use the Hamiltonian $H'_{op}(\epsilon)$. It is convenient to introduce new creation and annihilation operators through the equations

$$\hat{b}_0 = \hat{a}_0 - V^{\frac{1}{2}}\frac{\epsilon e^{-i\varphi}}{\mu} ; \quad \hat{b}_{\underline{k}} = \hat{a}_{\underline{k}} , \quad \underline{k} \neq 0 . \tag{2.19}$$

The operator $H'_{op}(\epsilon)$ then become

$$H'_{op}(\epsilon) = V\frac{\epsilon^2}{\mu} + \sum_{\underline{k}}[\epsilon(\underline{k})-\mu]\,\hat{b}_{\underline{k}}^+\hat{b}_{\underline{k}} , \tag{2.20}$$

and we have now a perfect gas of 'b-excitations'. The grand partition function now becomes

$$Z_{gr}(\epsilon) = \exp\left(-\frac{\beta V\epsilon^2}{\mu}\right) . Z_{gr}(0) , \tag{2.21}$$

and equations (2.11), (2.14), (2.15) and (2.16) become

$$n = \frac{\epsilon^2}{\mu^2} + \lim_{V\to\infty}\left[\frac{1}{V}\frac{z}{1-z} + \frac{1}{v_0}g_{\frac{3}{2}}(z)\right] , \tag{2.22}$$

$$x = \left[\frac{\beta^2\epsilon^2}{n-\zeta(\frac{3}{2})/v_0}\right]^{\frac{1}{2}} , \tag{2.23}$$

$$n_0 = \frac{\epsilon^2}{\mu^2} , \tag{2.24}$$

$$\langle(n-\langle n\rangle)^2\rangle_\epsilon = \lim_{V\to\infty}\left[\frac{\beta^2\epsilon^2}{V}\frac{1+z}{(1-z)^3} + \frac{1}{Vv_0}g_{\frac{1}{2}}(z)\right] , \tag{2.25}$$

and we see that now

$$\lim_{\epsilon\to 0}\langle(n-\langle n\rangle)^2\rangle_\epsilon = 0 . \tag{2.26}$$

Finally, we find for ψ:

$$\psi_{\varepsilon,\varphi} = \lim_{V\to\infty} \langle \psi_{op} \rangle_{\varepsilon,\varphi} = - \lim_{V\to\infty} \frac{1}{\beta V} \frac{\partial}{\partial(\varepsilon e^{i\varphi})} Z_{gr} = \frac{\varepsilon e^{-i\varphi}}{\mu} , \tag{2.27}$$

or using equation (2.24) ,

$$\psi = \lim_{\substack{\varepsilon\to 0 \\ \varphi \text{ fixed}}} \psi_{\varepsilon,\varphi} = \sqrt{n_0}\, e^{-i\varphi} . \tag{2.28}$$

We note that the average of ψ over all values of φ vanishes.

It is of some interest to look at the ground state of the Bose system described by $H'_{op}(\varepsilon)$. We saw that the transformation (2.19) reduced $H'_{op}(\varepsilon)$ essentially to $H'(0)$. This transformation can be written in the form

$$\hat{b}_{\underline{k}} = U_{op}(\varepsilon,\varphi)\, \hat{a}_{\underline{k}}\, U_{op}^{-1}(\varepsilon,\varphi) , \tag{2.29}$$

where

$$U_{op}(\varepsilon,\varphi) = \exp\left[-\tfrac{1}{2} V \frac{\varepsilon^2}{\mu^2}\right] \exp\left[V^{\frac{1}{2}} \frac{\varepsilon e^{-i\varphi}}{\mu} \hat{a}_0^{+}\right] \exp\left[- V^{\frac{1}{2}} \frac{\varepsilon e^{i\varphi}}{\mu} \hat{a}_0\right] . \tag{2.30}$$

The ground state, $|\phi_0(\varepsilon,\varphi)\rangle$, of $H'_{op}(\varepsilon)$ is the vacuum state for b-excitations, or

$$\hat{b}_0 |\phi_0(\varepsilon,\varphi)\rangle = 0 , \tag{2.31}$$

from which it follows that in the a-representation

$$\begin{aligned} |\phi_0(\varepsilon,\varphi\rangle &= U_{op}(\varepsilon,\varphi)|\text{vacuum}\rangle \\ &= \exp\left(-\tfrac{1}{2} V \frac{\varepsilon^2}{\mu^2}\right) \sum_N \frac{V^{N/2} (\varepsilon e^{-i\varphi}/\mu)^N}{\sqrt{N!}} |N\rangle , \end{aligned} \tag{2.32}$$

where $|N\rangle$ is the state with N bosons, all in the zero-momentum state:

$$|\text{vacuum}\rangle = (a_0^{+})^N |\text{vacuum}\rangle . \tag{2.33}$$

The state (2.32) satisfies the equation

$$\hat{a}_0 |\phi_0(\varepsilon,\varphi)\rangle = V^{\frac{1}{2}} \frac{\varepsilon e^{-i\varphi}}{\mu} |\phi_0(\varepsilon,\varphi)\rangle , \tag{2.34}$$

as follows from equations (2.31) and (2.19). It is one of Glauber's coherent states.

At $T = 0$, when all particles are in the zero-momentum state we have (see equation (2.24))

$$\frac{\varepsilon^2}{\mu^2} = n_0 = n , \tag{2.35}$$

and in the limit as $\epsilon \to 0$, we find the ground state wave function

$$|\phi_0(\varphi)\rangle = \lim_{\substack{\epsilon \to 0 \\ \varphi \text{ fixed}}} |\phi_0(\epsilon, \varphi)\rangle = e^{-\frac{1}{2}\bar{N}} \sum_N \frac{\bar{N}^{N/2}}{\sqrt{N!}} e^{-i\varphi N} |N\rangle \quad , \tag{2.36}$$

where $\bar{N} = nV$.

We note that this state — which is produced by the Bogolyubov procedure — corresponds to a particle number distribution

$$P_{|\phi_0(\varphi)\rangle}(N) = e^{-\bar{N}} \frac{\bar{N}^N}{N!} \quad , \tag{2.37}$$

which has a maximum at $N = \bar{N}$ and which has the fluctuations described by (2.26), as can also directly be seen from the relations

$$\langle (n - \langle n \rangle)^2 \rangle = \lim_{V \to \infty} \frac{1}{V^2} \left\{ \langle N_{op}^2 \rangle - \langle N_{op} \rangle^2 \right.$$

$$= \lim_{V \to \infty} \frac{1}{V^2} \left\{ \langle \phi_0(\varphi) | \hat{a}_0^+ \hat{a}_0 \hat{a}_0^+ \hat{a}_0 | \phi_0(\varphi) \rangle - \langle \phi_0(\varphi) | \hat{a}_0^+ \hat{a}_0 | \phi_0(\varphi) \rangle^2 \right\}$$

$$= \lim_{V \to \infty} \frac{1}{V^2} \frac{V \epsilon^2}{\mu^2} = \lim_{V \to \infty} \frac{n}{V} = 0 \quad . \tag{2.38}$$

On the other hand, the grand-canonical particle distribution follows from equation (1.28) of Chapter 1 after a few manipulations:

$$P_{gr}(N) = \frac{1}{\bar{N}} e^{-N/\bar{N}} \quad , \tag{2.39}$$

with a maximum at $N = 0$ and fluctuations given by equation (1.1).

CHAPTER 5

The Equation of State of an Imperfect Gas

5.1 A Hard-sphere Gas with Attractive Forces

Let us consider a classical gas of hard spheres which attract one another weakly. Such a gas is governed by the Hamiltonian

$$H = \sum_i \frac{p_i^2}{2m} + \sum_i F(\underline{r}_i) + \frac{1}{2} \sum_{i \neq j} \varphi(\underline{r}_{ij}) , \qquad (1.1)$$

where

$$\begin{aligned} F(\underline{r}_i) &= 0 , \text{ if } \underline{r}_i \text{ lies within the volume } V ; \\ &= \infty , \text{ otherwise} \end{aligned} \qquad (1.2)$$

and

$$\begin{aligned} \varphi(\underline{r}_{ij}) &= \infty \ , \quad r_{ij} \leq a \ ; \\ \varphi(\underline{r}_{ij}) &< 0 \ , \quad r_{ij} > a \ , \end{aligned} \qquad (1.3)$$

$\underline{r}_{ij} = \underline{r}_i - \underline{r}_j \ , \quad r_{ij} = |\underline{r}_{ij}|$.

Ornstein in his Leiden thesis showed how one can use the Hamiltonian (1.1) to find the equation of state. We use equations (1.10) and (1.12) of Chapter 1 to write the partition function $Z_\Gamma^{(N)}$ in the form

$$Z_\Gamma^{(N)} = \frac{1}{v_0^N N!} \int \cdots \int \prod_i d^3\underline{r}_i \exp\left[-\tfrac{1}{2}\beta \sum_{i \neq j} \varphi_r(\underline{r}_{ij}) - \tfrac{1}{2}\beta \sum_{i \neq j} \varphi_a(\underline{r}_{ij}) \right] , \qquad (1.4)$$

where the integration over the $\underline{r}_i$ extends over the volume V because of equation (1.2), and where we have separated φ into a repulsive part φ_r and an attractive part φ_a. Let S(x) be a function defined by the equation (see equation (1.8) of Chapter 2)

$$S(x) = \theta(x - a) \ , \quad x \neq a \ ; \quad S(a) = 0 \ . \qquad (1.5)$$

We can then write

$$\exp\left[-\tfrac{1}{2}\beta \sum_{i \neq j} \varphi_r(\underline{r}_{ij}) \right] = \prod_{i < j} S(r_{ij}) \ . \qquad (1.6)$$

To deal with the attractive part we write

$$\frac{1}{2} \sum_{i \neq j} \varphi_a(\underline{r}_{ij}) \approx \frac{1}{2} \sum_i \left(\frac{N}{V} d^3\underline{r} \right) \varphi_a(\underline{r} - \underline{r}_i) = - \frac{1}{2} \frac{N^2 C}{V} , \tag{1.7}$$

where

$$C = - \int \varphi_a(\underline{r}) \, d^3\underline{r} \tag{1.8}$$

We get thus from equation (1.4)

$$Z_\Gamma^{(N)} = \frac{D}{v_o^N N!} \exp\left(\frac{1}{2} \frac{\beta C N^2}{V} \right) , \tag{1.9}$$

where

$$D = \int \cdots \int \left(\prod_i d^3\underline{r}_i \right) \prod_{i<j} S(\underline{r}_{ij}) . \tag{1.10}$$

If $V \gg N v_1$, where v_1 is the volume of one of the hard spheres, $v_1 = \frac{4\pi}{3} a^3$, D is approximately equal to V^N, while if V approaches $N v_1$, we would expect D to behave more or less as $(V - N v_1)$. We shall interpolate this behaviour by writing

$$D = (V - b)^N , \tag{1.11}$$

where b will be of the order of $N v_1$. We thus have

$$e^{-\beta F} = Z_\Gamma^{(N)} = \frac{1}{v_o^N N!} (V - b)^N \exp\left(\tfrac{1}{2} \beta C \frac{N^2}{V} \right) , \tag{1.12}$$

while we get for the pressure

$$\beta P = - \frac{\partial \beta F}{\partial V} = - \frac{\beta C N^2}{2 V^2} + \frac{N}{V - b} , \tag{1.13}$$

or

$$\beta \left(P + \frac{a}{V^2} \right) (V - b) = N , \quad a = \tfrac{1}{2} C N^2 , \tag{1.14}$$

which is the van der Waals equation.

We know that, if $\beta < \beta_{cr} = 27 Nb/8a$, the isotherms given by equation (1.14) are well-behaved and everywhere satisfy the thermodynamic stability condition $\partial P / \partial V \leq 0$. However, if $\beta > \beta_{cr}$, the isotherm has the shape shown in Fig. 5.1, and one must supplement equation (1.14) with the Maxwell rule which leads to the isotherm with a horizontal part drawn under the equal area rule.

Fig.5.1

Let us briefly discuss two difficulties connected with Ornstein's derivation. First of all, we would expect that a correct evaluation of $Z_{\Gamma}^{(N)}$ should always lead to a stable isotherm without having to invoke Maxwell's rule. The reason why we are led to an isotherm with an unstable part is that we have (tacitly) assumed that we are dealing with a one-phase system. Secondly, the approximation (1.11) is known to be in error, because if we put $\varphi_a \equiv 0$ so that we are dealing with a hard-sphere gas, we can calculate the second and third virial coefficients exactly. Provided we put $b = 4Nv_1$, the second virial coefficient agrees with the one following from equation (1.14), but higher virial coefficients do not agree with their van der Waals values.

We shall now follow van Kampen and show how one can for a certain model derive equation (1.14) together with the Maxwell rule. This model involves a very-long-range attractive interaction between the particles — which makes it unnecessary to study in too much detail the hard-spheres geometry — and thus leaves open the fundamental question of why actual physical systems, for which the attractive forces are short-range, show phase transitions. However, the model clearly shows the essential feature of condensation which is the competition between low-energy liquid configurations and gas configurations corresponding to large volumes in phase space.

We start again from $Z_{\Gamma}^{(N)}$ in the form (1.10) of Chapter 1 and we can thus restrict our discussion to a consideration of the configurational partition function Q_N. In order to evaluate Q_N we divide the volume V into cells of size Δ, where Δ is so large that each cell contains a large number of particles, but at the same time so small that φ_a is practically constant within Δ. This means (i) that the range of φ_a must be large compared to the hard-sphere radius a, and (ii) that the density is sufficiently high so that many particles interact simultaneously.

A given configuration will now be characterized by a set of numbers N_i which give the number of particles in the ith cell, which is situated around the point $\underline{r}_i$. We now get for Q_N an expression similar to equations (1.4) or (1.9):

$$Q_N = \frac{1}{N!} \sum_{\{N_i\}} A(N_i)\ B(N_i)\ C(N_i) \quad , \tag{1.13a}$$

where $\{N_i\}$ indicates an N_i-configuration. The N_i must satisfy the condition

$$\sum_i N_i = N \quad , \tag{1.14a}$$

and the factors A , B and C in equation (1.13a) have the following meaning. The first factor is a combinatorial expression :

$$A(N_i) = \frac{N!}{\prod_i N_i!} \quad , \tag{1.15}$$

The factor $B(N_i)$ takes the repulsive part of the potential into account, so that by analogy with equation (1.10) and equation (1.11) we can to a first approximation write

$$B(N_i) = \prod_i \left[\int \cdots \int d^3\underline{r}_1 \ldots d^3\underline{r}_{N_1} \prod S \right]$$

$$\cong \prod_i (\Delta - N_i\delta)^{N_i} \tag{1.16}$$

where we shall assume δ to be a constant of the order of v_1. Finally, if we use the fact that φ_a is assumed to be constant over the dimensions of a cell, we can write $C(N_i)$ in the form

$$C(N_i) = \exp\left[-\tfrac{1}{2}\beta \sum_{i,j} N_i N_j \varphi_{ij} \right], \quad \varphi_{ij} = \varphi_a(\underline{r}_i - \underline{r}_j) \ . \tag{1.17}$$

If we use the fact that according to our assumptions the N_i are large numbers so that we can use Stirling's formula for the factorials, and if we write

$$\Phi(N_i) = \sum_i \left[N_i \ln(\Delta - N_i\delta) - N_i \ln N_i + N_i \right] - \tfrac{1}{2}\beta \sum_{i,j} N_i N_j \varphi_{ij} \ , \tag{1.18}$$

we have altogether for Q_N the expression

$$Q_N = \sum_{\{N_i\}} \exp \Phi(N_i) \ . \tag{1.19}$$

The next step is to find the equilibrium configuration, that is, the largest term in equation (1.19), satisfying equation (1.14a). Put differently, we must find the <u>absolute</u> maximum of equation (1.18). Taking equation (1.14a) into account in the usual way by means of a Langrangian multiplier, α, which afterwards can be found by substitution into equation (1.14a), we find that the N_i must satisfy the equations

$$\ln \frac{\Delta - N_i\delta}{N_i} - \frac{N_i\delta}{\Delta - N_i\delta} - \beta \sum_j \varphi_{ij} N_j = \alpha \ . \tag{1.20}$$

Let us first of all look for a solution of equation (1.20) corresponding to a uniform density, that is a solution of the form

$$N_i = \frac{N\Delta}{V} = n\Delta \; . \qquad (1.21)$$

If we write

$$\varphi_0 = \sum_i \varphi_{ij}\,\Delta = \int \varphi_a(\underline{r})\; d^3\underline{r} \quad , \qquad (1.22)$$

and use units in which δ is unity, we see first of all that in those units n must lie between 0 and 1, and that it must satisfy the equation

$$\ln\frac{1-n}{n} - \frac{1}{1-n} - \beta\varphi_0\, n = \alpha \quad , \qquad (1.23)$$

while the corresponding, extremum, value of Φ is

$$\Phi_{extr} = Vf(n) \quad , \qquad (1.24)$$

with

$$f(n) = n\ln\frac{1-n}{n} + n - \tfrac{1}{2}\beta\varphi_0\, n^2 \; . \qquad (1.25)$$

If Φ_{extr} is an absolute maximum, we can replace the sum in equation (1.19) by its largest term, $\exp\Phi_{extr}$, so that — apart from unimportant additive terms which do not depend on the volume — the free energy is given by

$$\beta F = -\,\Phi_{extr} \; . \qquad (1.26)$$

Restoring δ for a moment, we now get

$$\beta P = \frac{n}{1-n\delta} + \tfrac{1}{2}\beta\varphi_0\, n^2 \quad , \qquad (1.27)$$

or

$$\beta\left(P - \frac{\varphi_0 N^2}{2V^2}\right)(V - N\delta) = N \quad , \qquad (1.28)$$

which is the same as equation (1.14). To some extent we can say that so far all we have done is somewhat better to justify the use of equation (1.7).

Let us note that equation (1.23) can be written in terms of $f(n)$ as follows

$$f'(n) = \alpha \quad , \qquad (1.29)$$

and let us now investigate whether the solution (1.21) corresponds to a maximum of Φ. To investigate this we consider the matrix of the second derivatives of Φ:

$$\frac{\partial^2\Phi}{\partial n_i\,\partial n_j} = -\,\delta_{ij}\,\frac{\Delta^2}{N_i(\Delta - N_i)^2} - \beta\varphi_{ij} \; . \qquad (1.30)$$

We now prove a mathematical lemma. If for all i

$$a_i > \sum_j b_{ij} , \tag{1.31}$$

the real symmetric matrix

$$A_{ij} = a_i \delta_{ij} - b_{ij} , \quad b_{ij} = b_{ji} \geq 0 , \tag{1.32}$$

will be positive definite. This can be seen as follows. If λ is an eigenvalue of A_{ij} and x_i the corresponding eigenvector, we have

$$\lambda x_i = \sum_j A_{ij} x_j = a_i x_i - \sum_j b_{ij} x_j . \tag{1.33}$$

If now x_1 is the component of x_i with the largest absolute magnitude, and if we choose the arbitrary phase factor in x_i such that $x_1 > 0$, we have

$$(a_1 - \lambda) x_1 = \sum_j b_{ij} x_j \leq x_1 \sum_j b_{1j} < x_1 a_1 , \tag{1.34}$$

so that all eigenvalues are positive.

If, on the other hand, for all i

$$a_i \leq \sum_j b_{ij} , \tag{1.35}$$

we see that for the vector $y_i \equiv 1, 1, 1, \ldots ,$ we have

$$\sum_{ij} y_i A_{ij} y_j \leq 0 . \tag{1.36}$$

Hence, the matrix equation (1.30) is negative definite, provided

$$\frac{\Delta^2}{N_i(\Delta - N_i)^2} > - \sum_i \beta \varphi_{ij} . \tag{1.37}$$

Hence, using equations (1.21) and (1.22) we have

$$\frac{1}{n(1-n)^2} + \beta \varphi_0 > 0 , \tag{1.38}$$

or, using equation (1.25),

$$f''(n) < 0 . \tag{1.39}$$

The uniform density solution thus corresponds to a maximum as long as the $f'(n)$-curve has a negative slope. As the maximum value of $1/n(1-n)^2$ is 27/4, it follows that if the temperature satisfies the condition

$$-\beta \varphi_0 < \frac{27}{4} , \tag{1.40}$$

inequality (1.39) is always satisfied. Moreover, by the same argument, for any configuration $\{N_i\}$ the matrix of the second derivatives $\partial^2\Phi/\partial N_i\,\partial N_j$ is negative definite, if equation (1.40) is satisfied, so that Φ can have only one maximum and that must, therefore, be the one we have already found. The stable state of the system at temperatures above the critical temperature T_{cr} determined by the equation

$$T_{cr} = \frac{-4\,\varphi_o}{27\,k_B} , \tag{1.41}$$

thus corresponds to the equation of state (1.28).

Let us now consider temperatures such that

$$-\beta\varphi_o > \frac{27}{4} . \tag{1.42}$$

Now $f'(n)$ is no longer a monotonically decreasing function of n, but has a hump (see Fig. 5.2). There is now a range of α-values for which equation (1.23) has three solutions: n_I, n_{II} and n_{III}. The first two correspond to equation (1.39) and the corresponding value of Φ is at least a relative maximum, but n_{III} corresponds to a positive value of $f''(n)$, and hence to condition (1.35), which means that the corresponding value of Φ is not a maximum and thus does not correspond to a physical state.

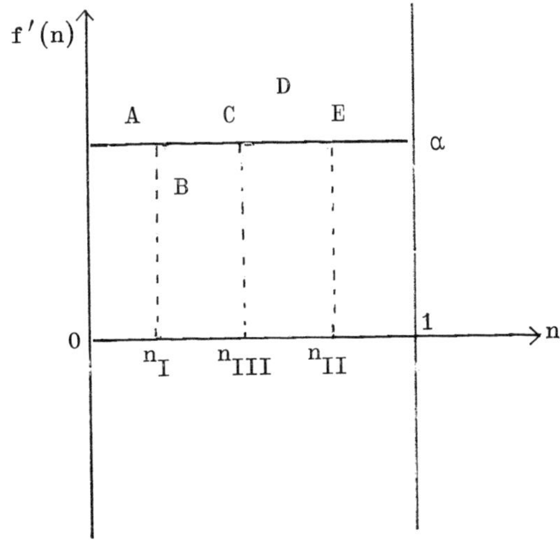

Fig. 5.2

We can now interpret the densities n_I and n_{II} as the densities of possibly coexisting phases. If there are, however, two coexisting phases, we are no longer dealing with the uniform case, and we can no longer use equation (1.21). We must thus go back to equation (1.20). Instead of equation (1.21) we now put

$$N_i = n(\underline{r}_i)\,\Delta , \tag{1.43}$$

and we get from equation (1.18)

$$\Phi = \int n(\underline{r}) \left[\ln \frac{1-n(\underline{r})}{n(\underline{r})} + 1 \right] d^3\underline{r} - \tfrac{1}{2}\beta \iint \varphi_a(\underline{r}-\underline{r}')\, n(\underline{r})\, n(\underline{r}')\, d^3\underline{r}\, d^3\underline{r}' , \tag{1.44}$$

or, if we use equation (1.25) ,

$$\Phi = \int f[n(\underline{r})]\, d^3r + \tfrac{1}{4}\beta \iint \varphi_a(\underline{r}-\underline{r}')\, [n(\underline{r}) - n(\underline{r}')]^2\, d^3\underline{r}\, d^3\underline{r}' \quad . \qquad (1.45)$$

Instead of equation (1.20) we now get

$$\frac{\delta\Phi}{\delta n(\underline{r})} \equiv f'[n(\underline{r})] + \beta \int \varphi_a(\underline{r}-\underline{r}')\, [n(\underline{r}) - n(\underline{r}')]\, d^3\underline{r}' = \alpha \quad . \qquad (1.46)$$

Let us assume that $n(\underline{r})$ varies so slowly that we can write

$$\varphi_a[n(\underline{r}) - n(\underline{r}')] \approx -\frac{1}{6}\varphi_a |r'-r|^2 \nabla^2 n(\underline{r}) \quad . \qquad (1.47)$$

From equation (1.46) we then get

$$f'[n(\underline{r})] - \tfrac{1}{2}\beta\varphi_2 \nabla^2 n(\underline{r}) = \alpha \quad , \qquad (1.48)$$

where

$$\varphi_2 = \tfrac{1}{3}\int r^2 \varphi_a(r)\, d^3\underline{r} \quad . \qquad (1.49)$$

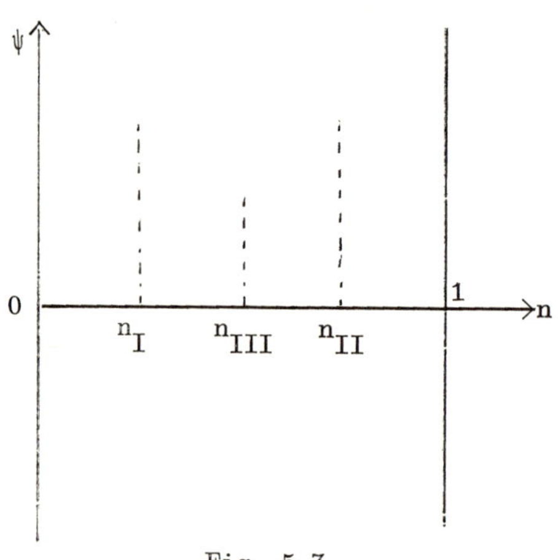

Fig. 5.3

Let us next assume that $n(r)$ depends on x only, so that equation (1.48) becomes

$$-\tfrac{1}{2}\beta\varphi_2 \frac{d^2n}{dx^2} = -f'(n) + \alpha \quad . \qquad (1.50)$$

We can now obtain all we need from this equation by noting that it is the classical equation of motion of a point particle of 'mass' $-\frac{1}{2}\beta\varphi_2$, with 'position' n at 'time' x moving in a potential $\psi(n)$ given by (see Fig. 5.3)

$$\psi(n) = f(n) - \alpha n = n \ln\frac{1+n}{n} + (1-\alpha)n - \tfrac{1}{2}\beta\varphi_0 n^2 \quad . \qquad (1.51)$$

Possible solutions are clearly once again $n = n_I$, n_{II}, or n_{III}, but these are the uniform density solutions in which we are no longer interested. We shall instead look for a solution such that

$$n \to n_I \quad \text{as} \quad x \to -\infty; \quad n \to n_{II} \quad \text{as} \quad x \to +\infty \quad . \qquad (1.52)$$

Looking at Fig. 5.3 and remembering the analogy with the classical motion, this means that we look for the case where the particle starts on one hill-top and ends up on the other hill-top. This can clearly only be realized for that value of α , say α_0 , for which

$$\psi(n_I) = \psi(n_{II}) \quad , \tag{1.53}$$

or

$$\int_{n_I}^{n_{II}} f'(n)\, dn = \alpha_0 (n_{II} - n_I) \quad , \tag{1.54}$$

which in terms of Fig. 5.2 means that the area ABC is equal to the area CDE : a Maxwell construction, though not yet in the P–V diagram. That it corresponds to the equal area construction in the P–V diagram can be seen as follows. For the van der Waals curve P satisfies the equation

$$\beta P = \frac{\partial \Phi}{\partial V} = f(n) - nf'(n) \quad , \tag{1.55}$$

or from equation (1.29) for $n = n_I$ or n_{II} :

$$\beta P = f(n) - \alpha n \, . \tag{1.56}$$

From equations (1.51) and (1.53) it then follows that

$$P_I = P_{II} \, . \tag{1.57}$$

Using equations (1.55), (1.54) and (1.57) we then have

$$\int_{II}^{I} P_I \, dV = \frac{N}{\beta} \int_{I}^{II} \frac{f - nf'}{n^2} \, dn = \frac{N}{\beta} \frac{f}{n} \Bigg|_{I}^{II} = P(V_I - V_{II}) \quad , \tag{1.58}$$

which concludes the proof.

We now call $n_I : n_g$ and $n_{II} : n_\ell$, and assume that apart from a negligible part of the volume, where there is a transition from the gas phase (n_g) to the liquid phase (n_ℓ), the density is either n_g or n_ℓ. Let V_g and V_ℓ be the volumes filled with densities n_g and n_ℓ. We then have

$$V_g + V_\ell = V \quad . \tag{1.59}$$

There is thus a set of two-phase states, all with $\alpha = \alpha_0$. As clearly

$$N = V_g n_g + V_\ell n_\ell \quad , \tag{1.60}$$

we see that these states occur for all average densities $n (= N/V)$ in the interval

$$n_g \leq n \leq n_\ell \quad . \tag{1.61}$$

It is instructive to note that n_g and n_ℓ must satisfy equations (1,29) and (1.53), or

$$f'(n_g) = f'(n_\ell) \quad , \tag{1.62}$$

$$f(n_g) - n_g\, f'(n_g) = f(n_\ell) - n_\ell\, f'(n_\ell) \ . \tag{1.63}$$

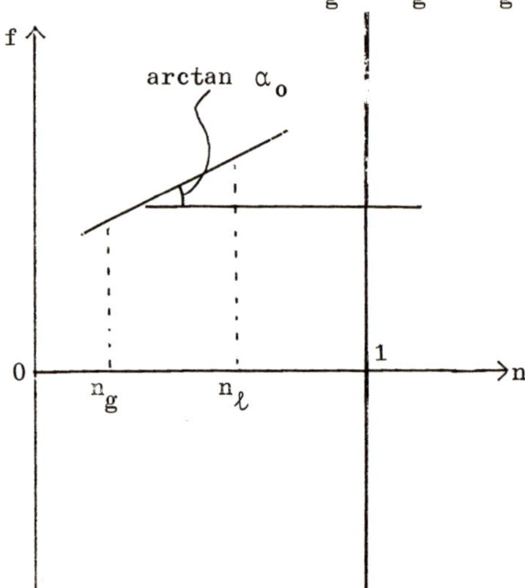

Fig. 5.4

which means that in a f- n-plot (see Fig. 5.4) these densities are found by the double-tangent construction, where the slope of the tangent is α_0.

As long as we may neglect the interface free energy, we have from equation (1.45)

$$-\beta F = \Phi = V_g \Phi(n_g) + V_\ell \Phi(n_\ell) \ , \tag{1.64}$$

and thus for the pressure

$$\beta P = \frac{\partial \Phi}{\partial V} = \frac{\partial V_g}{\partial V} P_g + \frac{\partial V_\ell}{\partial V} P_\ell = P_g (= P_\ell) \tag{1.65}$$

Finally, we note that the free energy corresponding to the horizontal part of the isotherm is lower than the free energy on the van der Waals curve. This follows from Fig. 5.4, as the double tangent lies above the curve, or

$$V_g \Phi(n_g) + V_\ell \Phi(n_\ell) > V\,\Phi\left(\frac{V_g}{V}\, n_g + \frac{V_\ell}{V}\, n_\ell\right) \tag{1.66}$$

5.2 A One-dimensional Model of a van der Waals Gas

We shall now discuss a one-dimensional mode, due to Kac and studied in detail by Kac, Hemmer and Uhlenbeck, which can be treated exactly and which shows a phase transition.

The model consists of N particles moving in a one-dimensional 'volume' L interacting through the equivalent of the potential (1.3) which we now make more specific as follows

$$\varphi = \infty, \ 0 \leq x \leq a \ ; \quad \varphi = -\eta e^{-\gamma x}, \ a \leq x \ . \tag{2.1}$$

Writing

$$\nu = \beta \eta \ , \tag{2.2}$$

and

$$\Lambda = \sqrt{\frac{2\pi\beta\hbar^2}{m}} \tag{2.3}$$

so that $\Lambda^3 = v_0$ (compare equation (1.11) of Chapter 1), we have the following expression for the partition function:

$$Z_\Gamma^{(N)} = \frac{1}{\Lambda^N}\frac{1}{N!}\int_0^L \dots \int_0^L dx_1 \dots dx_N \exp\left[\nu \sum_{i<j} e^{-\gamma|x_i - x_j|}\right] \prod_{i<j} S(|x_i - x_j|) , \tag{2.4}$$

where we have used equation (1.5). As the integrand is symmetric in the x_i and as the hard cores impose a linear ordering on the particles, we can rewrite equation (2.4) in the form

$$Z_\Gamma^{(N)} = \frac{e^{-\frac{1}{2}N\nu}}{\Lambda^N} \int_{0 < x_1 < x_2 \dots < x_N < L} \dots \int dx_1 \dots dx_N$$

$$\times \exp\left[\tfrac{1}{2}\nu \sum_{i,j} e^{-\gamma|x_i - x_j|}\right] \prod_{j=1}^{N-1} S(x_{j+1} - x_j) \quad . \tag{2.5}$$

Using the identity

$$\exp\left[-\tfrac{1}{2}\sum a_{k\ell} z_k z_\ell\right]$$

$$= \frac{1}{(2\pi)^{n/2}} \frac{1}{\sqrt{\text{Det } a}} \int_{-\infty}^{+\infty} \dots \int_{-\infty}^{+\infty} \prod_j dy_j \exp\left[-\tfrac{1}{2}\sum_{k,\ell} A_{k\ell} y_k y_\ell + i \sum_k y_k z_k\right] , \tag{2.6}$$

when the $A_{k\ell}$ are the matrix elements of the inverse of the matrix $a_{k\ell}$, we can write

$$\exp\left[\tfrac{1}{2}\nu \sum_{i,j} e^{-\gamma|x_i - x_j|}\right]$$

$$= \int_{-\infty}^{+\infty} \dots \int_{-\infty}^{+\infty} dy_1 \dots dy_N \; e^{\sqrt{\nu}(y_1 + \dots + y_N)} \; W(y_1) \prod_{j=1}^{N} P(y_j | y_{j+1} , x_{j+1} - x_j) , \tag{2.7}$$

where

$$W(y) = \frac{1}{\sqrt{(2\pi)}} e^{-\frac{1}{2}y^2} , \tag{2.8}$$

$$P(y|z,x) = \frac{1}{\sqrt{\{2\pi(1 - e^{-2\gamma x})\}}} \exp\left[\frac{-(z - y\, e^{-\gamma x})^2}{2(1 - e^{-2\gamma x})}\right] . \tag{2.9}$$

We now Laplace transform $Z_\Gamma^{(N)}$ with respect to L, and as after introducing equation (5.7) the x_i only occur in the combination $x_j - x_{j-1}$, we find

$$\int_0^\infty e^{-sL} Z_\Gamma^{(N)} dL = \int_0^\infty dL\, e^{-sL} \int_0^L dx_1 \int_{x_1}^L dx_2 \ldots \int_{x_{N-1}}^L dx_N \prod_{j=1}^{N-1} F_j(x_{j+1} - x_j)$$

$$= \int_0^\infty dx_1 \int_{x_1}^\infty dx_2 \ldots \int_{x_{N-1}}^\infty dx_N \int_{x_N}^\infty dL\, e^{-sL} \prod_{j=1}^{N-1} F_j(x_{j+1} - x_j)$$

$$= \frac{1}{s^2} \prod_{j=1}^{N-1} \int_a^\infty d\tau\, e^{-s\tau} F_j(\tau) , \tag{2.10}$$

where

$$\tau_1 = x_1 ,\quad \tau_2 = x_2 - x_1 , \ldots ,\quad \tau_j = x_j - x_{j-1} , \ldots ,\quad \tau_{N+1} = L - x_N , \tag{2.11}$$

and where the integration over τ_1 and τ_{N+1} led to the factor s^{-2}.

If we introduce the function

$$p_s(y|z) = \int_a^\infty d\tau\, e^{-s\tau} P(y|z,\tau) \tag{2.12}$$

we find that

$$\int_0^\infty e^{-sL} Z_\Gamma^{(N)} dL = \frac{e^{-\frac{1}{2}N\nu}}{\Lambda^N s^2} \int_{-\infty}^{+\infty} \ldots \int_{-\infty}^{+\infty} dy_1 \ldots dy_N\, e^{\sqrt{\nu}(y_1 + \ldots + y_N)} W(y_1) \prod_{j=1}^{N-1} p_s(y_j | y_{j+1}) ,$$

or

$$\int_0^\infty e^{-sL} Z_\Gamma^{(N)} dL = \frac{e^{-\frac{1}{2}N\nu}}{\Lambda^N s^2} \int_{-\infty}^{+\infty} \ldots \int_{-\infty}^{+\infty} dy_1 \ldots dy_N\, e^{\frac{1}{2}\sqrt{\nu}(y_2 + y_N)} \tag{2.13}$$

$$\times \sqrt{\{W(y_1)\, W(y_N)\}} \prod_{j=1}^{N-1} K_s(y_j , y_{j+1}) , \tag{2.14}$$

where $K_s(y,z)$ is given by the equation

$$K_s(y,z) = \frac{W(y) p_s(y|z)}{\sqrt{\{W(y)\, W(z)\}}}\, e^{\frac{1}{2}\sqrt{\nu}(y+z)} . \tag{2.15}$$

Consider now the Kac integral equation which has $K_s(y,z)$ as its symmetric kernel:

$$\int_{-\infty}^{+\infty} \psi(z)\, K_s(y,z)\, dz = \lambda\, \psi(y) . \tag{2.16}$$

One can prove that

$$\iint K_s(y,z)\,\psi(y)\,\psi(z)\,dy\,dz \geq 0 \,, \tag{2.17}$$

and

$$\iint K_s^2(y,z)\,dy\,dz < \infty \,, \tag{2.18}$$

which entails that equation (2.16) has a discrete set of positive eigenvalues $\lambda_i(s)$. If we order those such that

$$\lambda_o(s) > \lambda_1(s) > \lambda_2(s) \ldots \,,$$

$\lambda_i(s)$ will tend to zero as i tends to ∞; the corresponding eigenfunctions $\psi_i(y,s)$ will form a complete orthonormal set, and the kernel can be expanded in a convergent series:

$$K_s(y,z) = \sum_i \lambda_i(s)\,\psi_i(y,s)\,\psi_i(z,s) \,. \tag{2.19}$$

Substituting equation (2.19) into (2.14), and using the orthonormality of the ψ_i, we can integrate over $y_2, y_3, \ldots, y_{N+1}$, and are left with

$$\int_0^\infty e^{-sL}\, Z_\Gamma^{(N)}\, dL = \frac{e^{-\frac{1}{2}\nu N}}{\Lambda^N s^2} \sum_{j=0} \lambda_j^{N-1}(s)\, A_j^2 \,, \tag{2.20}$$

where

$$A_j = \int_{-\infty}^{+\infty} dy\; \psi_j(y,s)\sqrt{W(y)}\; e^{\frac{1}{2}y\sqrt{\nu}} \,. \tag{2.21}$$

Let us now turn to the grand partition function. From equation (1.15) of Chapter 1 it follows that

$$\begin{aligned}\int_0^\infty e^{-sL}\, e^q\, dL &= \int_0^\infty e^{-sL}\, dL \sum_N h^{3N}\, e^{\alpha N}\, Z_\Gamma^{(N)} \\ &= \sum_{j=0} \frac{A_j^2}{s^2 \lambda_j(s)} \sum_{N=1} \left[\frac{h^3 \lambda_j\, e^{\alpha-\frac{1}{2}\nu}}{\Lambda}\right]^N \\ &= \sum_{j=0}^\infty \frac{h^3 e^{\alpha-\frac{1}{2}\nu}}{\Lambda s^2}\, \frac{A_j^2}{1 - h^3 \lambda_j\, e^{\alpha-\frac{1}{2}\nu}\, \Lambda^{-1}} \,.\end{aligned} \tag{2.22}$$

From equation (2.22) it follows that the radius of convergence of the Laplace transform of e^q is that value of s for which

$$h^3 \lambda_o\, e^{\alpha-\frac{1}{2}\nu} = \Lambda \,. \tag{2.23}$$

On the other hand, the radius of convergence is also q/L. Moreover, we have for a homogeneous system (compare (1.19) of Chapter 1; L now takes the place of V)

$$\beta PL = q \,, \tag{2.24}$$

and we find thus

$$\lim_{L\to\infty} \frac{q}{L} = \beta P = s_o \,, \tag{2.25}$$

where s_0 is the solution of equation (2.23). Hence, we have

$$\lambda_0(\beta P) = \Lambda e^{\frac{1}{2}\nu-\alpha}/h^3 \quad . \tag{2.26}$$

As α/β is the partial Gibbs free energy per particle, we have from the thermodynamic relation $V = N(\partial G/\partial P)_T$:

$$\frac{L}{N} = \ell = \left(\frac{\partial(\alpha/\beta)}{\partial P}\right)_\beta \quad , \tag{2.27}$$

or, using equation (2.26),

$$\ell = - \left.\frac{\lambda_0'(s)}{\lambda_0(s)}\right|_{s=\beta P} \quad . \tag{2.28}$$

One can prove (i) that $\lambda_0(s)$ is a non-degenerate eigenvalue of equation (2.16), (ii) that $\lambda_0(s)$ is an analytic function of s, and (iii) that $\lambda_0(s)$ satisfies the inequality

$$\lambda_0'(s)^2 - \lambda_0(s)\lambda_0''(s) < 0 \tag{2.29}$$

for all values of s .

These properties mean that ℓ is a monotonically decreasing and analytical function of βP: there is no phase transition. Note that equation (2.28) gives ℓ as function of P rather than P as function of ℓ for the equation of state.

We shall now study the so-called van der Waals limit, that is, the limit as $\gamma \to 0$, where we keep the ratio η/γ finite as follows :

$$\lim_{\gamma\to 0} \int_a^\infty \varphi \, dx = - \eta_0 \quad . \tag{2.30}$$

This means that we are dealing with infinitely weak attractive forces with an infinitely long range.

Corresponding to equation (2.30) and equation (2.2) we write

$$\nu_0 = \lim_{\gamma\to 0} \frac{\nu}{\gamma} = \lim_{\gamma\to 0} \frac{\beta\eta}{\gamma} = \beta\eta_0 \quad . \tag{2.31}$$

Consider now the following expression

$$\sum_i \lambda_i^2 = \iint_{-\infty}^{+\infty} dy_1 dy_2 K_s(y_1,y_2)\, K_s(y_2,y_1)$$

$$= \iint_{-\infty}^{+\infty} dy_1 dy_2 e^{-\sqrt{(\nu_0\gamma)}(y_1+y_2)} \iint_a^{\infty} d\tau_1 d\tau_2 e^{-s(\tau_1+\tau_2)} P(y_1|y_2,\tau_1)\, P(y_2|y_1\tau_2)$$

$$= \iint_a^{\infty} d\tau_1 d\tau_2\, e^{-s(\tau_1+\tau_2)} \times \iint_{-\infty}^{+\infty} dy_1 dy_2\, e^{-\sqrt{(\nu_0\gamma)}(y_1+y_2)} \frac{e^{-\frac{1}{2}\xi_1^2 - \frac{1}{2}\xi_2^2}}{2\pi\sqrt{\left[(1-e^{-2\gamma\tau_1})(1-e^{-2\gamma\tau_2})\right]}} \tag{2.32}$$

where

$$\xi_1 = \frac{y_2 - y_1 e^{-\gamma\tau_1}}{\sqrt{(1-e^{-2\gamma\tau_1})}}, \quad \xi_2 = \frac{y_1 - y_2 e^{-\gamma\tau_2}}{\sqrt{(1-e^{-2\gamma\tau_2})}}. \tag{2.33}$$

Changing from y_1, y_2 to ξ_1, ξ_2 as integration variables, and taking into account that the relevant Jacobian is given by the equation

$$\frac{\partial(\xi_1,\xi_2)}{\partial(y_1,y_2)} = \frac{1-e^{-\gamma(\tau_1+\tau_2)}}{\sqrt{\left[(1-e^{-2\gamma\tau_1})(1-e^{-2\gamma\tau_2})\right]}}, \tag{2.34}$$

we get from equation (2.32)

$$\sum_i \lambda_i^2 = \iint_a^{\infty} d\tau_1 d\tau_2\, e^{-s(\tau_1+\tau_2)} \times \iint_{-\infty}^{+\infty} d\xi_1 d\xi_2 \frac{\exp[-\frac{1}{2}\xi_1^2 - \frac{1}{2}\xi_2^2 - \sqrt{(\nu_0\gamma)}(A_1\xi_1 + A_2\xi_2)]}{2\pi(1-e^{-\gamma(\tau_1+\tau_2)})}, \tag{2.35}$$

where

$$\left.\begin{aligned} A_1 &= \left(1+e^{-\gamma\tau_2}\right)\frac{\sqrt{(1-e^{-2\gamma\tau_1})}}{1-e^{-\gamma(\tau_1+\tau_2)}} \\ A_2 &= \left(1+e^{-\gamma\tau_1}\right)\frac{\sqrt{(1-e^{-2\gamma\tau_2})}}{1-e^{-\gamma(\tau_1+\tau_2)}} \end{aligned}\right\} \tag{2.36}$$

We now take the limit as $\gamma \to 0$, and the result is

$$\lim_{\gamma\to 0} \gamma \sum_i \lambda_i^2 = \iint_a^{\infty} d\tau_1 d\tau_2 \; \frac{e^{-s(\tau_1+\tau_2)}}{2\pi(\tau_1+\tau_2)}$$
$$\times \iint_{-\infty}^{+\infty} d\xi_1 d\xi_2 \exp\left[-\tfrac{1}{2}\xi_1^2 - \tfrac{1}{2}\xi_2^2 - \frac{2\sqrt{(2\nu_0\tau_1)}\,\xi_1 + 2\sqrt{(2\nu_0\tau_2)}\,\xi_2}{\tau_1+\tau_2} \right] . \tag{2.37}$$

To simplify this equation we introduce a delta-function integral as follows

$$1 = \int_{-\infty}^{+\infty} d\psi\, \delta\left[\psi - \frac{\xi_1/\tau_1 + \xi_2\sqrt{\tau_2}}{\tau_1+\tau_2} \right]$$
$$= \frac{1}{2\pi} \int_{-\infty}^{+\infty} d\psi \int_{-\infty}^{+\infty} dv \exp\left[iv \left\{ \psi - \frac{\xi_1\sqrt{\tau_1} + \xi_2\sqrt{\tau_2}}{\tau_1+\tau_2} \right\} \right]$$
$$= \frac{\tau_1+\tau_2}{2\pi} \int_{-\infty}^{+\infty} d\psi \int_{-\infty}^{+\infty} dw \exp\left[iw\,\psi(\tau_1+\tau_2) - iw(\xi_1\sqrt{\tau_1} + \xi_2\sqrt{\tau_2}) \right] . \tag{2.38}$$

Substituting expression (2.38) into the integrand of (2.37), and interchanging the order of integration, we can integrate over ξ_1 and ξ_2, and we obtain

$$\lim_{\gamma\to 0} \gamma \sum_i \lambda_i^2 = \frac{1}{(2\pi)^2} \int_{-\infty}^{+\infty} d\psi \int_{-\infty}^{+\infty} dw \iint_a^{\infty} d\tau_1 d\tau_2 \; e^{(-s+iw\psi)(\tau_1+\tau_2)}$$
$$\times \iint_{-\infty}^{+\infty} d\xi_1 d\xi_2 \exp\left[-\tfrac{1}{2}\xi_1^2 - \tfrac{1}{2}\xi_2^2 - 2\sqrt{(2\nu_0)}\,\psi_0 - iw\,\xi_1\sqrt{\tau_1} - iw\xi\sqrt{\tau_2} \right]$$
$$= \frac{1}{2\pi} \int_{-\infty}^{+\infty} d\psi \int_{-\infty}^{+\infty} dw \left[\int_a^{\infty} d\tau\; e^{-s\tau + iw\psi\tau - \sqrt{(2\nu_0)}\,\psi - \frac{1}{2}w^2\tau} \right]^2 . \tag{2.39}$$

Extending this calculation to higher powers of λ_i, and writing $w = \chi - i\psi$, $\psi = -\psi$, we find

$$\lim_{\gamma\to 0} \gamma \sum_i \lambda_i^n = \frac{1}{2\pi} \int_{-\infty}^{+\infty} d\psi \int_{-\infty}^{+\infty} d\chi \left[f(\psi,\chi) \right]^n , \tag{2.40}$$

where

$$f(\psi,\chi) = e^{\sqrt{(2\nu_0)}\,\psi} \int_a^{\infty} d\tau \exp\left[-\left\{ s + \tfrac{1}{2}(\psi^2 + \chi^2) \right\} \tau \right] . \tag{2.41}$$

It is thus plausible to suggest that, in general

$$\lim_{\gamma\to 0} \gamma \sum_i g(\lambda_i) = \frac{1}{2\pi} \iint_{-\infty}^{+\infty} d\psi\, d\chi\, g\left[f(\psi,\chi) \right] . \tag{2.42}$$

Let us now for $g(x)$ take the step function

$$g(x) = 1 , \quad \rho < x < \sigma ; \qquad g(x) = 0 , \quad \text{otherwise} . \tag{2.43}$$

If $N_\gamma(\rho,\sigma)$ is the number of eigenvalues λ_i between ρ and σ, we have now from equation (2.42)

$$\lim_{\gamma\to 0} \gamma N_\gamma(\rho,\sigma) = \frac{1}{2\pi} \times \text{area in } \psi,\ \chi\text{-plane for which } \rho < f(\psi,\chi) < \sigma . \tag{2.44}$$

Let $f(\psi,\chi)$ have an absolute maximum $\omega(s)$. If we take $\rho = \omega - \epsilon$, $\sigma = \omega$ the right-hand side of equation (2.44) will be finite so that $N_\gamma(\omega-\epsilon,\omega)$ must be infinite. Hence, $\omega(s)$ must be a limit point, and thus must be the maximum eigenvalue $\lambda_0(s)$. We have thus

$$\lim_{\gamma\to 0} \lambda_0(s,\gamma) = \omega(s) = \operatorname*{Max}_{\psi,\chi} f(\psi,\chi) = \operatorname{Max} F(\psi) \tag{2.45}$$

where we have maximized $f(\psi,\chi)$ with respect to χ:

$$F(\psi) = f(\psi,0) = \frac{e^{-a(s+\frac{1}{2}\psi^2) + \psi\sqrt{(2\nu_0)}}}{s+\frac{1}{2}\psi^2} . \tag{2.46}$$

If we define a function $h(\psi)$ by the equation

$$h(\psi) = a\psi + \frac{\psi}{s+\frac{1}{2}\psi^2} , \tag{2.47}$$

the maximum of $F(\psi)$ occurs when s satisfies the equation

$$h(\psi) = \sqrt{(2\nu_0)} . \tag{2.48}$$

To see what this means let us first assume that

$$8\,sa > 1 . \tag{2.49}$$

In that case

$$\frac{dh}{d\psi} = a + \frac{1}{s+\frac{1}{2}\psi^2} - \frac{\psi^2}{(s+\frac{1}{2}\psi^2)^2} , \tag{2.50}$$

will always be positive, and equation (2.48) has, therefore, exactly one solution, $\psi(s)$, which substituted into equation (2.46) gives us $\lambda_0(s)$.

To find the equation of state, we note that equation (2.28) can be written in the form

$$\ell = \frac{h(\psi)}{\psi} , \tag{2.51}$$

or

$$\ell = \frac{\sqrt{(2\nu_0)}}{\psi} \tag{2.52}$$

where we have used equation (2.48).

Using the maximization condition (2.48) we can write equation (2.28) in the form

$$\ell = - \frac{\partial \ln F(\psi(s), s)}{\partial s} = - \frac{\partial \ln F}{\partial s} - \frac{\partial \ln F}{\partial \psi} \frac{\partial \psi}{\partial s}$$

$$= a + \frac{1}{s + \frac{1}{2}\psi^2} . \tag{2.53}$$

If we substitute equation (2.52) for ψ and equation (2.25) for s into equation (2.53), and use equation (2.31), we get

$$\beta\left(P + \frac{\eta_0}{\ell^2}\right)(\ell - a) = 1 . \tag{2.54}$$

This is the van der Waals equation. Note that η_0 is given by equation (2.30) while C in (1.14) was given by equation (1.8), the factor 2 arising from the fact that the integration in equation (2.30) is taken over positive values of x only.

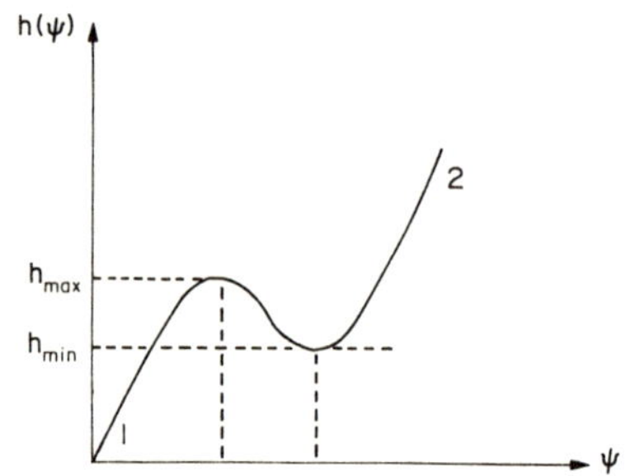

Fig. 5.5

We must now consider the case when

$$8\,sa < 1 . \tag{2.55}$$

In that case, $h(\psi)$ as function of behaves as shown in Fig. 5.5: there is a minimum, h_{min}, and a maximum, h_{max}. If

$$h_{min} > \sqrt{(2\nu_0)} , \tag{2.56}$$

or if

$$h_{max} < \sqrt{(2\nu_0)} , \tag{2.57}$$

equation (2.48) has one solution, corresponding again to the van der Waals equation (2.54).

However, if

$$h_{min} < \sqrt{(2\nu_0)} < h_{max} , \tag{2.58}$$

there are three roots, two corresponding to local maxima and one to a local minimum of $F(\psi)$. If for fixed ν_0 one varies s (which is similar to varying ν_0 for fixed s), there must be a value $\bar{s}(\nu_0)$ for which the two maxima are the same; this means that the local maximum which was the absolute maximum and which started, say, on branch 1 of Fig. 5.5 will have to give precedence to the local maximum on branch 2. From Fig. 5.5 we see that this means that $h(\psi)$ changes abruptly, which means – as we can see from equation (2.51) – that the isotherm will have a horizontal section. Of course, the 'liquid' and the 'gas' parts of the isotherm, corresponding, respectively, to the branches 1 and 2 of Fig. 5.5, satisfy the van der Waals' equation. Moreover, we see from equation (2.26) that as for $s = \bar{s}$ the two local maxima are equal, the Gibbs free energy is the same at both ends of the horizontal section, that is: the construction of this section is carried out in accordance with the thermodynamic Maxwell rule.

The critical point occurs when $dh/d\psi = 0$ has a double root, or

$$8\, s_{cr} a = 1 \, , \tag{2.59}$$

and from equation (2.50) it follows that then

$$s_{cr} + \tfrac{1}{2}\psi_{cr}^2 = \frac{1}{2a} \, . \tag{2.60}$$

From equations (2.60), (2.47), (2.51), (2.52), (2.59) and (2.54) we then find that

$$\ell_{cr} = 3a \, ; \quad \beta_{cr} = \frac{27a}{8\eta_0} \, , \quad P_{cr} = \frac{\eta_0}{27a^2} \, , \tag{2.61}$$

which are the usual van der Waals relations.

CHAPTER 6
A Simple Derivation of the Bloch Equation

In 1946 Bloch introduced the following equation of motion for the magnetization $\underline{M}$:

$$\frac{d\underline{M}}{dt} = \gamma[\underline{M}\times\underline{B}] - \frac{\underline{i}M_x + \underline{j}M_y}{T_2} - \frac{\underline{k}(M_z - M_0)}{T_1} . \qquad (1)$$

The applied magnetic field is the same here as in section 2.4: a constant field $\underline{B}_0$ along the z-axis and an oscillating field $\underline{B}_1$ in the x y-plane. In (1) $\underline{i}$, $\underline{j}$, and $\underline{k}$ are unit vectors along the x- , y-, and z-axes, γ is the magneto-gyric ratio ($\gamma = 2\mu_B/\hbar$ for a spin-$\frac{1}{2}$ particle), M_0 is the equilibrium magnetization in the field B_0, and T_1 and T_2 are, respectively, the spin-lattice or longitudinal and the spin-spin or transverse relaxation times.

Before deriving a simplified form of equation (1), let us give a simple phenomenological explanation of the various terms in equation (1). The system we shall be considering in this section is that of N spin-$\frac{1}{2}$ particles interacting with their surroundings (lattice or 'bath') and with the external magnetic field.

If $\underline{S}_i$ is the spin-vector of the ith spin, the total magnetization will be given by the equation

$$\underline{M} = \sum_i \gamma\hbar\underline{S}_i = N\gamma\hbar\langle\underline{S}\rangle , \qquad (2)$$

where the pointed brackets indicate some kind of average. If a field $\underline{B}$ is present, the spins will experience a torque, and if $\underline{\mu}_i$ is the magnetic moment of the ith spin, its classical equation of motion is

$$\frac{d\hbar\underline{S}_i}{dt} = [\underline{\mu}_i \times \underline{B}] , \qquad (3)$$

or

$$\frac{d\underline{M}}{dt} = \gamma[\underline{M}\times\underline{B}] , \qquad (4)$$

giving us the first term on the right-hand side of the Bloch equation. Ehrenfest's theorem about quantum-mechanical equations of motion then ensures that equation (4) also holds for quantum-mechanical systems where the averages are now quantum-statistical averages.

Let us now consider the term involving T_1. If a spin-$\frac{1}{2}$ particle is placed in a steady field $\underline{B}_0$, its spin energy will be $\pm\frac{1}{2}\gamma\hbar B_0$ depending on whether the spin is antiparallel or parallel to the field. If N_+ and N_- are the number of spins parallel or antiparallel to the field $(N_+ + N_- = N)$, at equilibrium we have

$$N_\pm^{eq} = N \frac{\exp(\pm\frac{1}{2}\beta\gamma\hbar B_0)}{2\cosh[\frac{1}{2}\beta\gamma\hbar B_0]} , \tag{5}$$

or, if $\frac{1}{2}\beta\gamma\hbar B_0 \ll 1$, as is practically always the case,

$$N_\pm^{eq} = \frac{1}{2}N[1 \pm \frac{1}{2}\beta\gamma\hbar B_0] . \tag{6}$$

In equation (5) and (6) β is again given by equation (1.2) of Chapter 1.

The magnetization of the system is given by the equation

$$M = \frac{1}{2}\gamma\hbar[N_+ - N_-] , \tag{7}$$

and we have thus

$$M_0 = \frac{1}{4}\beta\gamma^2\hbar B_0 , \tag{8}$$

Interaction with the surrounding will induce transitions from the up- to the down-spin position and vice versa. Let W_+ and W_- be the transition probabilities for an up (down) spin to become a down (up) spin. In that case we have

$$\frac{dN_+}{dt} = -\frac{dN_-}{dt} = W_-N_- - W_+N_+ , \tag{9}$$

whence follows the equation of motion for the magnetization:

$$\frac{dM}{dt} = \gamma\hbar(W_-N_- - W_+N_+) . \tag{10}$$

At equilibrium $dN_+/dt = 0$, whence it follows that

$$\frac{W_+}{W_-} = \frac{N_-^{eq}}{N_+^{eq}} , \tag{11}$$

or, if we use equation (6),

$$W_\pm = W(1 \mp \frac{1}{2}\beta\gamma\hbar B_0) , \tag{12}$$

where W is the average of W_+ and W_-.

From equations (8), (10) and (12) we now find

$$\frac{dM}{dt} = -\frac{M - M_0}{T_1} , \quad T_1 = \frac{1}{2W} . \tag{13}$$

To see the physical meaning of the terms involving T_2, we note (see equation (3)) that the magnetic moments of the individual spins will precess around the magnetic field which they experience. As there is no steady field in the xy-plane, there will be no non-vanishing equilibrium magnetization in that plane but, due to fluctuations, at some moment there might occur a resultant component in that plane. If all spins are precessing with the same frequency (γB) this resultant non-vanishing component would persist. However, apart from the field B_0 which is the same for all the spins there will be fluctuating fields B_{fl} due to the other spins. These fields will be much weaker than any applied fields. They will, however, fluctuate in both magnitude and direction and this will result in a spread of precession frequency $\Delta\omega$ of the order of γB_{fl}. Hence any temporarily non-vanishing magnetization component in the xy-plane will decay with a characteristic time T_2 of the order of $(\gamma B_{fl})^{-1}$.

Having seen the physical nature of the processes which lead to the relaxation of the magnetization we shall now study the behaviour of a spin-system described by the Hamiltonian

$$H_{op} = -\tfrac{1}{2}\gamma\hbar\left(\underline{\sigma}_{op}\cdot[\underline{B}_0 + \underline{B}_1 + \underline{B}_{fl}]\right) \tag{14}$$

where the components of $\underline{\sigma}_{op}$ are the Pauli matrices (compare equation (3.2) of Chapter 2)

$$\sigma_x = \begin{pmatrix} 0 & 1 \\ 1 & 0 \end{pmatrix}, \quad \sigma_y = \begin{pmatrix} 0 & -i \\ i & 0 \end{pmatrix}, \quad \sigma_z = \begin{pmatrix} 1 & 0 \\ 0 & -1 \end{pmatrix}, \tag{15}$$

where $\underline{B}_0$ is a steady field, $\underline{B}_1$ an oscillating field which has a time-dependence $e^{i\omega t}$, where $\omega \approx \omega_0 = \gamma B_0$ (that is, we consider the resonance region), and where we assume that $B_0 \gg B_1$. Finally $\underline{B}_{fl}$ is a fluctuating field. This can be expressed by stating that the average value of $\underline{B}_{fl}$ is zero, while at the same time we expect $\underline{B}_{fl}$ to 'remember' its original value over a period equal to the correlation time τ_c. Mathematically we express this by writing

$$\lim_{T\to\infty} \frac{1}{T}\int_t^{t+T} \underline{B}_{fl}(t')\,dt' = 0, \tag{16}$$

$$\lim_{T\to\infty} \frac{1}{T}\int_t^{t+T} dt' \int_t^{t'} dt''\, B^i_{fl}(t')\, B^j_{fl}(t'') = \delta_{ij}\, k, \qquad i,j = x,y,z, \tag{17}$$

where we expect k to be of the order of the product of τ_c and the square of the amplitude of the fluctuating field. The δ_{ij} factor in equation (17) expresses the independence of the different components of $\underline{B}_{fl}$.

We shall describe the system by a density matrix ρ_{op} and use equations (1.22) and (1.24) of Chapter 1 from which we get for the equation of motion of $\langle g\rangle$

$$i\hbar\frac{d}{dt}\langle g\rangle = \langle [g_{op},H_{op}]_-\rangle . \tag{18}$$

Let us use equation (18) to derive equation (4) for the case of a constant field $\underline{B}$, so that the Hamiltonian is simply

$$H_{op} = -\tfrac{1}{2}\gamma(\underline{\sigma}_{op},\underline{B}) . \tag{19}$$

From equations (18) and (2) we then get

$$\frac{d\underline{M}}{dt} = \tfrac{1}{2}N\gamma\hbar\frac{d\langle\underline{\sigma}_{op}\rangle}{dt} = \frac{N\gamma\hbar}{i\hbar}\langle[\underline{\sigma}_{op} - \tfrac{1}{2}\gamma\hbar(\underline{\sigma}_{op}.\underline{B})]_-\rangle , \tag{20}$$

and using the commutation relations of the components of $\underline{\sigma}_{op}$ (compare equation (6.4) of Chapter 2) which we symbolically write in the form

$$[\underline{\sigma}_{op}\times\underline{\sigma}_{op}] = 2i\,\underline{\sigma}_{op} , \tag{21}$$

we recover equation (4). We note by the way that this derivation is not restricted to the spin-$\frac{1}{2}$ case but is equally valid for general S .

Equation (1) describes an irreversible approach to equilibrium, and in order to derive it — or a simplified version of it — from the reversible equation (18) we must use one of the usual tricks of statistical mechanics. In this particular case we shall use a 'coarse graining' in time, which means that we write

$$\frac{d\langle\underline{\sigma}_{op}\rangle}{dt} = \frac{\langle\underline{\sigma}_{op}\rangle_{t+\Delta t} - \langle\underline{\sigma}_{op}\rangle_t}{\Delta t} = \frac{1}{\Delta t}\int_t^{t+\Delta t}\frac{d\langle\underline{\sigma}_{op}\rangle}{dt'}\,dt' . \tag{22}$$

Here Δt is a time interval which is sufficiently short so that the difference quotient is a good approximation to the differential quotient and which is sufficiently long so that we can use equations (16) and (17). The first condition means that Δt must satisfy the inequalities

$$\omega\Delta t \approx \omega_0\Delta t \ll 1 , \tag{23}$$

and the second condition that

$$\Delta t \gg \tau_c .$$

Combining (23) and (24) we see that it is necessary that

$$\omega_0\tau_c \ll 1 , \tag{25}$$

which is the extreme narrowing case. We shall assume equation (25) to be satisfied.

Let us now first see how ρ_{op} changes with time. From its equation of motion we find

$$i\hbar \frac{d\rho_{op}}{dt} = -\tfrac{1}{2}\gamma\hbar\left[\left(\underline{\sigma}_{op}\cdot[\underline{B}_0+\underline{B}_1+\underline{B}_{fl}]\right), \rho_{op}\right]_- , \qquad (26)$$

and neglecting $\underline{B}_1$ and $\underline{B}_{fl}$ in first approximation and integrating we get

$$\rho_{op}(t+\Delta t) = \exp\left[\tfrac{1}{2}i\,\gamma(\underline{\sigma}_{op}\cdot\underline{B}_0)\Delta t\right]\hat{\rho}(t)\exp\left[-\tfrac{1}{2}i\gamma(\underline{\sigma}_{op}\cdot\underline{B}_0)\Delta t\right]. \qquad (27)$$

As $\omega_0 = \gamma B_0$ we see that by virtue of equation (23), ρ_{op} does not change over the period Δt in this approximation. This conclusion would remain valid, if we had included $\underline{B}_1$ by virtue of the condition $B_0 \gg B_1$. Before looking into the effect of $\underline{B}_{fl}$, let us now look at the equation of motion for $\langle\underline{\sigma}_{op}\rangle$. We have from equations (18), (14) and (22)

$$\frac{d\langle\underline{\sigma}_{op}\rangle}{dt} = \frac{1}{\Delta t}\frac{-i}{\hbar}\int_t^{t+\Delta t} \mathrm{Tr}\ \rho_{op}(t')\left[\underline{\sigma}_{op} - \tfrac{1}{2}\gamma\hbar\left(\underline{\sigma}_{op}\cdot[\underline{B}_0+\underline{B}_1+\underline{B}_{fl}]\right)\right]_- dt' . \qquad (28)$$

As ρ_{op} to a first approximation is constant and as $\underline{B}_1$ in the same approximation is constant by virtue of equation (23), the terms involving $\underline{B}_0$ and $\underline{B}_1$ will lead to

$$\frac{d\langle\underline{\sigma}_{op}\rangle}{dt} = \gamma\left[\langle\underline{\sigma}_{op}\rangle\times[\underline{B}_0+\underline{B}_1]\right] = \gamma\left[\langle\underline{\sigma}_{op}\rangle\times\underline{B}\right] , \qquad (29)$$

by the same reasoning by which we derived equation (4) from equation (20). By virtue of equation (16) the term in $\underline{B}_{fl}$ does not contribute to this approximation.

Let us now see what happens if we go one step further in the evaluation of $\rho_{op}(t')$ from equation (26), taking into account the fact that $\underline{B}_0$ and $\underline{B}_1$ do not contribute to any change in ρ_{op}. We have

$$\rho_{op}(t') = \tfrac{1}{2}i\,\gamma\int_t^{t'}\left[\left(\underline{\sigma}_{op}\cdot\underline{B}_{fl}(t'')\right), \rho_{op}(t)\right]_- dt'' + \rho_{op}(t) , \qquad (30)$$

where we have used the fact that $\rho_{op}(t'')$ is a constant in first approximation to replace it by $\rho_{op}(t)$ in the integrand. Substituting equation (30) for $\rho_{op}(t')$ into the term involving $\underline{B}_{fl}(t')$ in equation (28), this term will become

$$\frac{i\gamma}{2\Delta t}\int_t^{t+\Delta t} dt'\ \mathrm{Tr}(\tfrac{1}{2}i\,\gamma)\int_t^{t'}\left[\left(\underline{\sigma}_{op}\cdot\underline{B}_{fl}(t'')\right), \rho_{op}(t)\right]_-\left[\underline{\sigma}_{op}, \left(\underline{\sigma}_{op}\cdot\underline{B}_{fl}(t')\right]_- dt''$$

$$= \frac{-\gamma^2}{4\Delta t}\int_t^{t+\Delta t} dt'\int_t^{t'} dt''\ \mathrm{Tr}\ \rho_{op}\left[\left[\underline{\sigma}_{op}\left(\underline{\sigma}_{op}\cdot\underline{B}_{fl}(t')\right)\right]_-, \left(\underline{\sigma}_{op}\cdot\underline{B}_{fl}(t'')\right)\right]_-$$

$$= -\,2\gamma^2 k\,\langle\underline{\sigma}_{op}\rangle , \qquad (31)$$

where we have used equations (21) and (17).

Combining equations (29) and (31) we get the simplified Bloch equation

$$\frac{d\underline{M}}{dt} = \gamma[\underline{M}\times\underline{B}] - \frac{\underline{M}}{T} \tag{32}$$

with

$$T = \frac{1}{2\gamma^2 k} \tag{33}$$

As $k \sim \tau_c |B_{fl}|^2$, we have

$$T = \frac{1}{2\gamma^2 k} \sim \frac{1}{\tau_c \gamma^2 |B_{fl}|^2} > \frac{1}{\tau_c \gamma^2 B_o^2} = \frac{\tau_c}{(\omega_o \tau_c)^2} \gg \tau_c \tag{34}$$

a condition which is, of course, necessary for the validity of our derivation.

Let us briefly discuss whether it would be possible to get by our procedure — without complicating it — the full Bloch equation (1) involving two relaxation times as well as a non-vanishing equilibrium magnetization $\underline{M}_o$.

As to the first point, we only considered one cause for relaxation, $\underline{B}_{fl}$. Of course, as the z-axis is a preferred direction we could have introduced on the right-hand side of equation (17) $\delta_{ij} k_i$ with $k_x = k_y \neq k_z$, but this would have been rather artificial.

The absence of $\underline{M}_o$ is at first sight more serious. We note, however, first of all, that we have described the influence of the surroundings by a classical field, that is, essentially considered the limit as $\hbar \to 0$, but from equation (8) this is equivalent to putting $\underline{M}_o = 0$. Moreover, we must bear in mind that the non-vanishing magnetization comes about because the spin system is kept at a constant temperature by the bath. When we derived equation (13) it was essential that there was a feedback from the spins to the lattice and as in our theory this feedback is of necessity absent, we cannot hope to recover $\underline{M}_o$. The only way to remedy this would be by introducing instead of the classical correlation functions (17) the quantum-mechanical ones and at the same time to take the spin-bath interaction into account in more detail. All this would, however, destroy the simplicity of the theory.

CHAPTER 7
Statistical Mechanics of Stellar Systems

7.1 Introduction

The first question we have to ask ourselves is whether it is at all possible to have a statistical mechanics of systems of gravitating mass points — which is an idealized description of stellar systems. Tacitly many people assume that, of course, any physical system can be described by the formalism of statistical mechanics which we have used in the preceding sections. After all, even a plasma can be so described, and Coulomb forces differ only in sign from gravitational forces. To that one must reply that the compensating background of oppositely charged particles which so conveniently is present and used in plasma physics is absent in the gravitational case. It is, however, true that the forces are now attractive. The difficulty lies in the fact that the gravitational potential $U(r)$ tends to $-\infty$ as r tends to zero. We can thus gain any arbitrary amount of energy by forming one close binary, and this energy can be used to let other members of the system escape to infinity: there are no walls to contain the system! The available phase space is infinite and as $U(r) \to 0$ as $r \to \infty$, the Maxwell-Boltzmann distribution does not have an exponential cut-off as it has in momentum space: we run into normalization difficulties.

One could, of course, hope that some kind of quasi-equilibrium can be established, and that the escape of stars from the system will be a slow process. In fact, the presence of star clusters which look like being in some kind of equilibrium points to this. It is thus of interest to study non-equilibrium processes, that is, the evolution of stellar systems. In recent years several numerical studies have been made. For instance, van Albada has studied the evolution of systems with N stars where N lies between 10 and 24. He observed the feature we described a moment ago: after a relatively short time — which was one or two orders of magnitude larger than the characteristic time scale of the system, which is the average time a star would take to cross the system, the system disintegrates and a more or less stable multiple system containing between $\frac{1}{3}$ to $\frac{1}{2}$ of the original group, and dominated by a close binary, with a binding energy of about $1\frac{1}{2}$ times the total (negative) energy of the group, remains. The disintegration process is essentially a redistribution of the binding energy among fewer and fewer stars. From his analysis it also follows that the important collisions are multiple ones rather than binary ones which throws some doubts on some calculations, such as the one at the beginning of section 2, of relaxation times. Before we draw any conclusions from van Albada's work, we must, however, point out that small N systems such as the ones considered by van Albada and large N systems such as star clusters show essential differences. Van Albada's work is, however, without doubt relevant to the question of the common occurrence of binary, triple and other multiple systems.

Let us consider a system of N stars of mass m and let R and v_{av} be the radius of the system and the average stellar velocity. Let us consider the ratio of the potential energy of a star due to another star at the average distance between stars in the system and its kinetic energy. The first quantity is $Gm^2/(R/N^{\frac{1}{3}})$ where G is the gravitational constant, and the second one is $\frac{1}{2}m\,v_{av}^2$. The ratio r is thus given by the equation

$$r \sim \frac{G\,m\,N^{\frac{1}{3}}}{R v_{av}^2} \tag{1.1}$$

For the total energy of the system we have

$$E_{tot} \approx \tfrac{1}{2}N\,m\,v_{av}^2 - \frac{N^2 m^2 G}{R}\,, \tag{1.2}$$

and if we invoke the virial theorem — which is only proved for closed systems — we have

$$-E_{tot} \approx \tfrac{1}{2}N\,m\,v_{av}^2\,, \quad -2\,E_{tot} \approx \frac{N^2 m^2 G}{R}\,, \tag{1.3}$$

so that we find from equation (1.1)

$$r \sim N^{-\frac{2}{3}}\ . \tag{1.4}$$

From this it follows that for large N , r is always small: the over-all gravitational field will be important, while for small N close encounters play an important rôle.

We can see the difference between small-N and large-N systems also by looking at the average energy of the relative motion of two stars in the system. We have

$$\langle E_{rel}\rangle \sim \langle\tfrac{1}{2}m\,v_{av}^2\rangle - \langle\frac{Gm^2}{d_{av}}\rangle \sim \frac{-E_{tot}}{N}\left[1 - \frac{4}{N^{\frac{2}{3}}}\right], \tag{1.5}$$

where d_{av} is the average distance between stars, and where once again we have invoked the virial theorem. We see that for large N most encounters are hyperbolic, but for small N we must expect elliptic motion, that is, (temporary) bound systems (binaries).

7.2 The Relaxation Time of Stellar Systems

Let us first of all consider the case where one star of mass m_2 collides with another star of mass m_1 and where the collision is characterized by an impact parameter p and an initial relative velocity V (see Fig. 7.1).

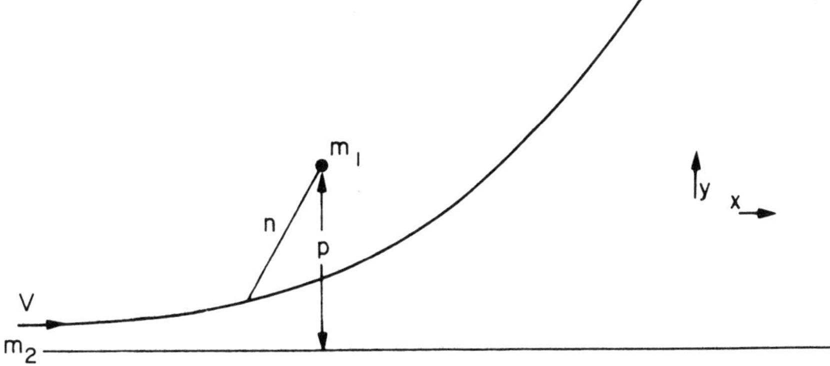

Fig. 7.1

If $\langle \Delta V^2 \rangle$ is the average of the square in the change in the relative velocity per unit time, we shall define a relaxation time $\tau_r^{(1)}$ by the equation

$$\tau_r^{(1)} = \frac{V^2}{\langle \Delta V^2 \rangle} . \tag{2.1}$$

To estimate $\langle \Delta V^2 \rangle$ we shall assume that the scattering angle is small so that we can equate the change in momentum ΔP during the collision to the change in its y-component. We have thus

$$\begin{aligned} \Delta P &\approx - \int_{-\infty}^{+\infty} F_y \, dt \\ &= - \int_{-\infty}^{+\infty} \frac{G m_1 m_2 y}{r^3} \, dt . \end{aligned} \tag{2.2}$$

As we are assuming the scattering angle to be small, we can write

$$r^2 \approx p^2 + V^2 t^2 , \quad y \approx p , \tag{2.3}$$

where we have taken $t = 0$ at the point of closest approach. From equations (2.2) and (2.3) we thus get

$$\Delta P \approx - \int_{-\infty}^{+\infty} \frac{G m_1 m_2 p}{(p^2 + V^2 t^2)^{\frac{3}{2}}} dt = \frac{2 G m_1 m_2}{p V} \quad . \qquad (2.4)$$

To find $\langle \Delta V \rangle^2$ we note that the number of collisions per unit time with impact parameter between p and dp and relative velocities lying within a velocity-space volume element $d^3\underline{V}$ is equal to $f(\underline{V}) d^3\underline{V} \, . \, V \, 2\pi p \, dp$. Using equations (2.4) and (2.1) we thus find

$$\frac{1}{\tau_r^{(1)}} \approx \frac{\iint V \, f(\underline{V}) \, p \, dp \, d^3\underline{V} \, G^2 m^2 / p^2 V^2}{V^2} \quad , \qquad (2.5)$$

or

$$\tau_r^{(1)} \approx \frac{V^2}{m^2 G^2} \frac{R^3}{N} \left(V^{-1} \right)_{av}^{-1} \ln^{-1} \frac{p_2}{p_1} \quad , \qquad (2.6)$$

where $(\ldots)_{av}$ indicates an average over the distribution function, where p_1 and p_2 are cut-off values of the impact parameter, indicating the limits outside which equation (2.4) loses its validity, and where the other quantities have the same meaning as in section 1.

For the upper limit we can take the size of the system,

$$p_2 \approx R \quad , \qquad (2.7)$$

and for the lower limit that impact parameter for which the scattering angle becomes of the order of unity, or put differently, the distance for which the ratio of the potential energy to the kinetic energy becomes of the order of unity (compare equation (1.1)), that is,

$$\frac{G m}{p_1 v_{av}^2} \approx 1 \quad . \qquad (2.8)$$

Using equation (1.3) we see that

$$p_1 \approx \frac{R}{N} \quad . \qquad (2.9)$$

Substituting these values for p_1 and p_2 into equation (2.6) and taking $m \approx 10^{33}$ g , $G \approx 10^{-7} \, g^{-1} \, cm^3 \, s^{-2}$, $N \sim 10^{11}$, $R \sim 10^{22}$ cm , $v \sim 10^7 \, cm \, s^{-1}$ which are typical values for a galaxy, we find

$$\tau_r^{(1)} \approx 10^{15} \text{ years} \quad , \qquad (2.10)$$

which is much longer than the usually accepted age of our universe. The question then arises why there are elliptical galaxies which from their light distribution look like having reached equilibrium. This is the question we shall now turn to. However, we want to draw attention to the fact that the equilibrium configuration does not in this case mean equipartition of energy as that would mean stars with smaller mass at the outside and this is not in agreement with the observed brightness distribution.

Taking our clue from what we saw in section 1 we shall look for relaxation due to the mean gravitational field following Lynden-Bell's treatment. Let ψ be the gravitational potential and let ϵ^* be the energy of a star of mass m, so that

$$\epsilon^* = m(\tfrac{1}{2}c^2 + \psi) , \tag{2.11}$$

where c is the velocity of the star. If ϵ is the energy per unit mass, $\epsilon = \epsilon^*/m$, we find that

$$\frac{d\epsilon^*}{dt} = m\frac{\partial\psi}{\partial t} , \tag{2.12}$$

or

$$\frac{d\epsilon}{dt} = \frac{\partial\psi}{\partial t} . \tag{2.13}$$

We now define a relaxation time $\tau_r^{(2)}$ by the equation

$$\tau_r^{(2)} = \left\langle\frac{(d\epsilon/dt)^2}{\epsilon^2}\right\rangle^{-\frac{1}{2}} , \tag{2.14}$$

or, if we use equation (2.13)

$$\tau_r^{(2)} = \left\langle\frac{(\partial\psi/\partial t)^2}{\epsilon^2}\right\rangle^{-\frac{1}{2}} . \tag{2.15}$$

Consider now the quantity

$$I = \sum m r^2 , \tag{2.16}$$

where the summation is over all the stars of the system, and where r is measured from the centre-of-mass. We find the following time-dependent virial theorem:

$$\tfrac{1}{2}\ddot{I} = \sum m(\underline{r}\cdot\ddot{\underline{r}}) + \sum m(\dot{\underline{r}}\cdot\dot{\underline{r}}) , \tag{2.17}$$

where a dot indicates a time derivative. Using Newton's equations of motion and Euler's relation for a homogeneous function, we get from equation (2.17)

$$\tfrac{1}{2}\ddot{I} = 2T + V , \tag{2.18}$$

where T is the total kinetic energy of the system — which is the sum of the kinetic energies of the individual stars — and V the total potential energy —

which is half the sum of the potential energies of the individual stars. We also have the total energy of the system E — which is an integral of motion — and which satisfies the equation

$$E = T + V \ . \tag{2.19}$$

At equilibrium $\ddot{I} = 0$ and we have (compare equations (1.2) and (1.3))

$$T = -E \ , \quad V = 2E \ . \tag{2.20}$$

If the system is not in equilibrium it may oscillate turning kinetic into potential energy and vice versa. As E is constant, and T is the sum of the contribution from single stars and V half the sum of such contributions, we find that on average

$$\tfrac{1}{2} m c^2 \sim - \tfrac{1}{4} m \psi \ , \tag{2.21}$$

or

$$\epsilon \sim \tfrac{3}{4} \psi \ , \tag{2.22}$$

from which we find, using equation (2.15), that

$$\tau_r^{(2)} = \tfrac{3}{4} \left\langle \frac{\dot{\psi}^2}{\psi^2} \right\rangle^{-\frac{1}{2}} \ . \tag{2.23}$$

As $\ln \psi$ will change on a time scale which is roughly that for an oscillation of the system, we expect $\tau_r^{(2)}$ to be of the order of the period of such an oscillation. To find its value approximately we note that if R is again the radius of the system we have approximately

$$V \approx - \frac{GM^2}{R} \ , \tag{2.24}$$

$$I \approx MR^2 \ , \tag{2.25}$$

where

$$M = Nm \ . \tag{2.26}$$

Equations (2.18), (2.19), (2.24) and (2.25) then lead to

$$\ddot{R}^2 = \frac{2E}{M} + \frac{GM}{R} \ . \tag{2.27}$$

For small amplitude oscillations $(R = R_o + \delta R \ , \ R_o = -GM^2/2E \ ,$ from equations (2.24) and (2.20)) we find

$$R_o \, \delta\ddot{R} \approx - \frac{GM}{R_o^3} \, \delta R \ , \tag{2.28}$$

and thus

$$\tau_r^{(2)} \approx \frac{R_o^{\frac{3}{2}}}{\sqrt{(GM)}} \ . \tag{2.29}$$

Let us compare $\tau_r^{(1)}$ and $\tau_r^{(2)}$. From equations (2.29), (2.26), (2.6), (2.7) and (2.9) we find

$$\frac{\tau_r^{(1)}}{\tau_r^{(2)}} \approx \frac{V^3 R^{\frac{3}{2}} \sqrt{(G N m)}}{m^2 G^2 N (\ln N)} . \tag{2.30}$$

If we now use $T \approx N m V^2 \approx G N^2 m^2/R$ (compare equations (2.24) and (2.20)), we finally find

$$\frac{\tau_r^{(1)}}{\tau_r^{(2)}} \approx \frac{N}{\ln N} . \tag{2.31}$$

The relaxation is in fact extremely fast, even though the oscillations are Landau damped. In the next subsection we shall discuss this Landau damping and in the subsection after that we shall discuss what form of equilibrium is the result of this 'mean field relaxation'.

7.3 Landau Damping

The Landau damping in a stellar system is completely analogous to that in a plasma. We shall present here a simplified — one-dimensional — account of the effect and will just note that in a star system any synchronism which may exist originally and which is manifested in the oscillations will be spoiled by the different periods of the stars around the centre of the system— this is completely equivalent to the different velocities which are present in a Maxwell distribution — which is the case we shall consider.

Let there be a one-dimensional density wave of the form

$$\rho = \rho_0 \, e^{i(\omega t - kx)} . \tag{3.1}$$

Let us work in a system of reference moving with the phase velocity of this wave, ω/k. In that frame of reference we have a standing wave, and the same is true for the gravitational potential which is related to ρ through the Poisson equation

$$\nabla^2 \psi = 4\pi G \rho . \tag{3.2}$$

Consider now a star which is moving with a velocity close to ω/k. In the frame in which we are working it sees a sinusoidal potential field (see Fig. 7.2). If at a certain time it is at x_0 it will be moving in a potential well. It will only be able to escape from this well if its kinetic energy in the moving frame of reference is sufficiently large, that is, if its velocity c in the moving frame exceeds a critical velocity c_{cr} which is essentially

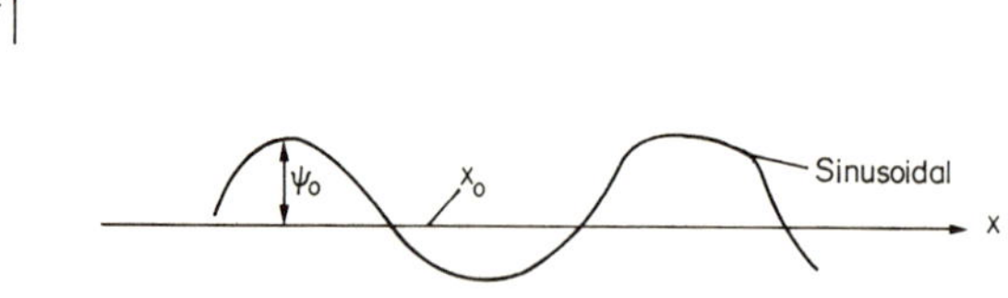

Fig. 7.2

given by the equation

$$\tfrac{1}{2}c_{cr}^2 = 2\psi_0 \quad , \tag{3.3}$$

where ψ_0 is the amplitude of the ψ-wave (see Fig. 7.2).

This means that all stars with velocities between $(\omega/k) - c_{cr}$ and $(\omega/k) + c_{cr}$ will be trapped and be forced to move with the phase velocity of the wave, ω/k. This in turn entails that stars with velocities between $(\omega/k) - c_{cr}$ and ω/k are accelerated and those with velocities between ω/k and $(\omega/k) + c_{cr}$ slowed down. As the velocity distribution $f(c)$ of the stars is likely to be a monotonically decreasing function of c (see Fig. 7.3) we see that on balance the wave loses energy to the stellar system: it is damped.

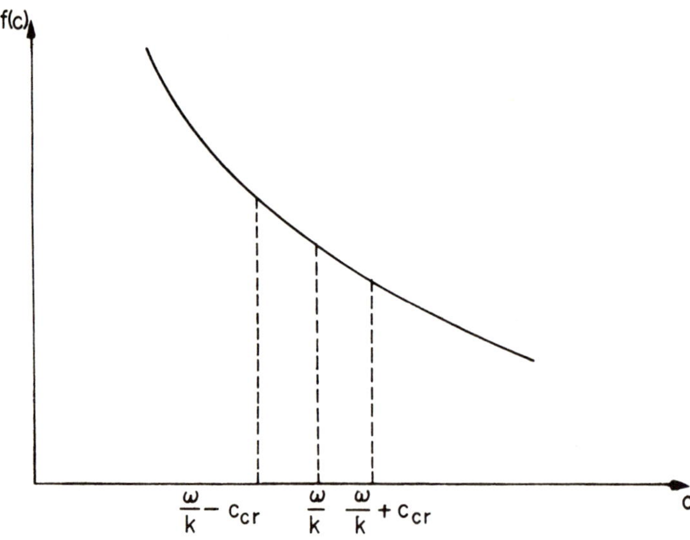

Fig. 7.3

7.4 Lynden-Bell Statistics

We now want to find out the distribution function which is reached through the violent relaxation. Let $f\, d^3\underline{r}\, d^3\underline{v}$ be the total mass of stars in a volume element of the 6-dimensional μ-space (I neglect here the difference between $\underline{r}$, $\underline{v}$-space and $\underline{r}$, $\underline{p}$-space). We know that collisions are spaced approximately at time intervals of the order of $\tau_r^{(1)}$ so that we can neglect them. The equation of motion for f is thus given by the equation

$$\frac{\partial f}{\partial t} + (\underline{v} \cdot \nabla) f + \left(\dot{\underline{v}} \cdot \frac{\partial}{\partial \underline{v}}\right) \underline{f} = 0 , \qquad (4.1)$$

which is a Liouville equation in six dimensions. The acceleration $\dot{\underline{v}}$ is determined by the gravitational potential,

$$\dot{\underline{v}} = - \nabla\psi , \qquad (4.2)$$

and ψ depends on f :

$$\psi(\underline{r}) = - G \int \frac{f(\underline{r}', \underline{v}')}{|\underline{r} - \underline{r}'|} d^3v' \, d^3r' . \qquad (4.3)$$

From these equations we see that (4.1) is a Vlasov equation.

As we have a Liouville equation in μ-space rather than in Γ-space, we have conservation of phase elements in μ-space. Put differently, the total number of elements of phase with a given f is conserved, or: we cannot have any overlap of elements of phase. If we now try to find f by the so-called elementary method, where we divide μ-space into micro-cells, we have a new kind of statistics (see Table) with distinguishable entities, satisfying an exclusion principle.

Table of Statistics

	Indistinguishability	Distinguishability
No exclusion	I. Bose-Einstein	II. Maxwell-Boltzmann
Exclusion	III. Fermi-Dirac	IV. Lynden-Bell

To find f we must first of all find the weight W of a macro-state characterized by having N_i 'particles' or 'elements of phase' in a macro-cell with Z_i micro-cells. We remember that we have for the Maxwell-Boltzmann case

$$W_{MB} = \frac{N!}{\prod_i N_i!} \prod_i Z_i^{N_i} , \qquad (4.4)$$

and for the Fermi-Dirac case

$$W_{FD} = \prod_i \frac{Z_i!}{(Z_i - N_i)!\, N_i!} . \qquad (4.5)$$

For the Lynden-Bell case we find

$$W_{LB} = N! \prod_i \frac{Z_i!}{(Z_i - N_i)!\,N_i!} , \tag{4.6}$$

which differs merely by a factor N! from the Fermi-Dirac case. As we know that there are very few aspects of statistical mechanics where the N! is important, we expect to find essentially a Fermi-Dirac distribution.

The subsidiary conditions are

$$E_{tot} = \text{const} = \int f \tfrac{1}{2} v^2 \, d^3\underline{v} \, d^3\underline{r} - \tfrac{1}{2} G \iint \frac{f f'}{|\underline{r} - \underline{r}'|} d^3\underline{r} \, d^3\underline{r}' \, d^3\underline{v} \, d^3\underline{v}' , \tag{4.7}$$

and

$$M_{tot} = \text{const} = \int f \, d^3\underline{v} \, d^3\underline{r} , \tag{4.8}$$

and in this way we find

$$f = \eta \left[e^{\beta(\epsilon - \mu)} + 1 \right]^{-1} , \tag{4.9}$$

where ϵ is the energy per unit mass, given by the equation (compare equation (2.11))

$$\epsilon = \tfrac{1}{2} c^2 + \psi , \tag{4.10}$$

which shows that we have (if $\eta/f \ll 1$, which is the usual case) a Maxwell-Boltzmann distribution with respect to ϵ, as observed, and not with respect to ϵ^*, as is usually the case

A few remarks are needed here. First of all, we note that Saslaw has derived equation (4.9) by considering the full Liouville equation and coarse-graining in time. He shows that if $U\{f\}$ is the mean gravitational potential

$$U\{f\} = C \int \frac{\int f(r_N', p_N') \, d p_N' \, (d r_N' / d\underline{r}_\alpha')}{|\underline{r} - \underline{r}_\alpha|} d^3\underline{r}_\alpha' , \tag{4.11}$$

where r_N' and p_N' stand for the 3N coordinate and momentum variables, and where $f(r_N', p_N')$ is the distribution function in Γ-space, violent relaxation will occur, provided we may write

$$\left\langle \left(\frac{\partial U\{f\}}{\partial \underline{r}} \cdot \frac{\partial f}{\partial \underline{p}} \right) \right\rangle \approx \left(\frac{\partial \langle U \rangle}{\partial \underline{r}} \cdot \frac{\partial \langle f \rangle}{\partial \underline{p}} \right) , \tag{4.12}$$

where pointed brackets indicate coarse-grained time averages (compare equation (22) of Chapter 6).

Secondly, we note that stellar systems have a negative specific heat. Because of the virial theorem (compare equations (1.2) and (1.3)), if we have two stellar systems which are approximately in equilibrium and which can exchange energy, the one which loses energy gets hotter, and the one which gains energy gets hotter. Similar processes, which may look like phase transitions, may happen within one system. A plausible conclusion to draw from this is that it is unlikely that the universe will tend to equilibrium, contrary to the view expressed, for instance, by Boltzmann.

Finally, we want to mention that numerical studies of large-N systems of gravitating mass-points have by and large confirmed Lynden-Bell's analysis.

7.5 Gravitational Polarization

In this last section we shall discuss briefly some work by Gilbert who started from the Liouville equation to describe relaxation effects in stellar systems. His work is similar to that by Balescu and by Lenard on plasmas. As the Balescu-Lenard treatment assumes the plasma to be uniform over distances of the order of the Debye length and as the equivalent of the Debye length for a stellar system is its radius, one has to proceed slightly differently.

Let

$$D(1, \ldots, N)\ d(1)\ d(2) \ldots d(N) \tag{5.1}$$

be the normalized probability of finding the position of star number 1 within $d^3\underline{r}_1$, its velocity within $d^3\underline{v}_1$, and its mass lying between m_1 and dm_1, where

$$d(1) \equiv d^3\underline{r}_1\ d^3\underline{v}_1\ dm_1 , \tag{5.2}$$

of finding the position, velocity, and mass of star 2 within $d(2)$, and so on. The normalization of D means that

$$\int D\ d(1)\ d(2) \ldots d(N) = 1 . \tag{5.3}$$

The Liouville equation, which is the equation of motion for D, is of the form

$$\frac{\partial D}{\partial t} + \sum_i (\underline{v}_i \cdot \nabla_i) D + \sum_{i \neq j} \left(\underline{A}_{ij} \cdot \frac{\partial}{\partial \underline{v}_j}\right) D = 0 , \tag{5.4}$$

where $\underline{A}_{ij}$ is the acceleration of the ith star due to the jth one,

$$A_{ij} = m_j G \frac{\underline{r}_j - \underline{r}_i}{|\underline{r}_j - \underline{r}_i|^3} = \frac{1}{N} \underline{a}_{ij} .$$

We have introduced the quantities $\underline{a}_{ij}$ in order to make all dependence on the number of stars, N, explicit, as our treatment will involve an expansion in inverse powers of N.

Let

$$f^{(n)}(1,\ldots,n) = \int D(1,\ldots,N)\, d(n+1)\ldots d(N) \tag{5.6}$$

be a reduced distribution function. The probability to find n stars, within the volume element $d(1),\ldots,d(n)$ will then be $\binom{N}{n} f^{(n)}$. For example, $Nf^{(1)}$ will be the probability density.

If $\psi(i)$ and $\varphi(i,j)$ are functions depending on the variables of a single star and of a pair of stars, respectively, we find for their averages

$$\langle \psi(i) \rangle = N \int \psi(1)\, f^{(1)}(1)\, d(1)\ , \tag{5.7}$$

and

$$\Big\langle \sum_{i<j} \varphi(i,j) \Big\rangle = \tfrac{1}{2} N(N-1) \int \varphi(1,2)\, f^{(2)}(1,2)\, d(1)\, d(2)\ . \tag{5.8}$$

More convenient for our purpose than D is the Bogolyubov generating functional

$$\mathcal{L}\{u\} = \int \prod_{\ell=1}^{N} \left[1 + \frac{u(\ell)}{N}\right] D(1,\ldots,N)\, d(1),\ldots,d(N)\ . \tag{5.9}$$

The $f^{(n)}$ follow from $\mathcal{L}$ from the equation

$$\left.\frac{\delta^n \mathcal{L}}{\delta u(1),\ldots,\delta u(n)}\right|_{u\equiv 0} = \frac{N^{-n}\, N!}{(N-n)!}\, f^{(n)}(1,\ldots,n)\ . \tag{5.10}$$

From the Liouville equation (5.4) it follows that $\mathcal{L}$ satisfies the equation

$$\frac{\partial \mathcal{L}}{\partial t} + \int u(1)(\underline{v}_1 \cdot \nabla) \frac{\delta \mathcal{L}}{\delta u(1)}\, d(1)$$

$$+ \iint u(1) \left[1 + \frac{u(2)}{N}\right]\left(\underline{a}_{12} \cdot \frac{\partial}{\partial \underline{v}_1}\right) \frac{\delta \mathcal{L}}{\delta u(1)\, \delta u(2)}\, d(1)\, d(2) = 0\ . \tag{5.11}$$

One can prove that a functional satisfying the boundary conditions following from the normalization of D and following from the requirement that as $N \to \infty$ the probability distributions for the different stars are independent is of the form

$$\mathcal{L} = \exp\left\{\int u(1)\, f(1)\, d(1) + \frac{1}{2N}\iint \left[g(1,2) - f(1)\, f(2)\right] u(1)\, u(2)\, d(1)\, d(2)\ . \right. \tag{5.12}$$

Expression (5.12) can be made to solve equation (5.11) to first order in N^{-1} by a suitable choice of f and g. Moreover, from equation (5.10) it follows that

$$f^{(1)}(1) = f(1) , \tag{5.13}$$

and

$$f^{(2)}(1,2) = f(1)\, f(2) + \frac{1}{N} g(1,2) . \tag{5.14}$$

Expression (5.12) is a solution — to first order in N^{-1} — of equation (5.11) provided f and g satisfy the equations

$$\frac{\partial f(1)}{\partial t} + \hat{H}_1 f(1) + \frac{1}{N}\left(\frac{\partial}{\partial \underline{v}_1} \cdot \int \underline{a}_{12}\, g(1,2)\, d(2)\right) - \frac{1}{N}\left(\underline{A}_1 \cdot \frac{\partial}{\partial \underline{v}_1}\right) f(1) = 0 , \tag{5.15}$$

and

$$\frac{\partial g(1,2)}{\partial t} + (\hat{H}_1' + \hat{H}_2')\, g(1,2) + \left[\left(\underline{a}_{12}^{*} \cdot \frac{\partial}{\partial \underline{v}_1}\right) + \left(\underline{a}_{21}^{*} \cdot \frac{\partial}{\partial \underline{v}_2}\right)\right] f(1)\, f(2) = 0 , \tag{5.16}$$

where

$$\underline{A}_1 = \int \underline{a}_{12}\, f(2)\, d(2) \tag{5.17}$$

that is, $\underline{A}_1$ is the mean gravitational acceleration at $\underline{r}_1$,

$$\underline{a}_{12}^{*} = \frac{\underline{a}_{12} - \underline{A}_1}{N} , \tag{5.18}$$

that is, $\underline{a}_{12}^{*}$ is the additional acceleration of star 1 when star 2 is known to be at $\underline{r}_2$ and certainly is nowhere else. The operators $\hat{H}$ and $\hat{H}'$ are defined by the equations

$$\hat{H}_1\, \psi(1) = (\underline{v}_1 \cdot \nabla_1)\, \psi(1) + \left(\underline{A}_1 \cdot \frac{\partial}{\partial \underline{v}_1}\right) \psi(1) , \tag{5.19}$$

and

$$\hat{H}_1'\, \psi(1) = \hat{H}_1 \psi(1) + \left(\frac{\partial f(1)}{\partial \underline{v}_1} \cdot \int \underline{a}_{12}\, \psi(2)\, d(2)\right) . \tag{5.20}$$

For a stationary state, to lowest order in N^{-1}, f must satisfy the equation

$$\hat{H} f = 0 , \tag{5.21}$$

and in equation (5.16) we can substitute for f solutions of equation (5.21).

Let us now look at some consequences of our equations. If we fix the position (or orbit) of one particular star which we shall call the test star and which we shall assume to be star 2, the effect upon an arbitrary star, star 1, in the system will be that its contribution to $\underline{A}_1$, which is $\underline{A}_1/N$ must be replaced by the definite acceleration $\underline{a}_{12}/N$, or: selecting a test star is equivalent to applying the extra acceleration $\underline{a}^{*}_{12}$.

If $f(1)$ is the unperturbed distribution function and $f^{*}(1|2)/N$ the perturbation resulting from selecting star 2 as test star, $f^{*}(1|2)$ which we can call the <u>polarization cloud</u> will satisfy the linearized Vlasov equation

$$\frac{\partial f^{*}(1|2)}{\partial t} + \hat{H}_1' f^{*}(1|2) + \left(\underline{a}^{*}_{12} \cdot \frac{\partial}{\partial \underline{v}_1}\right) f(1) = 0 . \tag{5.22}$$

One can express the correlation function $g(1,2)$ (compare equation (5.14)) in terms of $f^{*}(1|2)$ as follows

$$g(1,2) = f^{*}(1|2)\, f(2) + f^{*}(2|1)\, f(1) + \int f^{*}(1|3)\, f^{*}(2|3)\, f(3)\, d(3) \quad , \tag{5.23}$$

an equation similar to the one derived by Rostoker for a fully ionized plasma.

Having obtained $g(1,2)$ one can now write down a kinetic equation, but for this we refer to Gilbert's paper. Gilbert has solved his equations numerically for the case where a test star is situated at the centre of an isochrone cluster, and he finds that gravitational polarization may amplify the effective gravitational mass of a test star by a factor of 2.75 (the bare mass is augmented rather than shielded !). Clearly these kinds of collective effects may be important.

CHAPTER 8
Approach to Equilibrium

8.1 A Simple Model

Lamb has recently discussed a simple model which shows the main features of the approach to equilibrium. He believes that the essential feature of such an approach is the interaction of the system with a thermal reservoir (compare the section on the Bloch equation). As the system Lamb takes a quantum mechanical simple harmonic oscillator with coordinate E, frequency ω, eigenvalues $\hbar W_n$ $(= (n+\frac{1}{2})\hbar\omega)$, and eigenfunctions $h_n(E)$.

Let the system at $t=0$ be described by the wavefunction

$$\Psi(E) = \sum_{n=0}^{\infty} c_n h_n(E) \quad , \tag{1.1}$$

so that the probability P_n to find the state n is given by the equation

$$P_n = |c_n|^2 \quad . \tag{1.2}$$

If we use the density matrix ρ_{op} to describe the system, we have for the pure case

$$\rho_{nm} = c_m^* c_n \quad , \tag{1.3}$$

and $P_n = \rho_{nn}$, but for the case of a mixture we have

$$\rho_{op} = \sum_i w_i \rho_{op}[\Psi_i] \quad , \tag{1.4}$$

where w_i is the weight of the state Ψ_i and where all w_i are ≥ 0, and $\Sigma w_i = 1$.

The reservoir is assumed to consist of quantum-mechanical two-level atoms with energies $\hbar W_a$ and $\hbar W_b (<\hbar W_a)$ such that (exact resonance)

$$\omega = W_a - W_b \quad . \tag{1.5}$$

It is assumed that at any one time only one of the reservoir atoms is coupled to the system, and that the coupling lasts for a time τ. Let $u_{a,b}(x)$ be the eigenfunctions of the reservoir atoms, and let the atom at time $t=0$ be in the state u_b. At $t = 0$, the wavefunction of the combined oscillator-reservoir atom system will be

$$\Psi(E,x\,;\,0) = \Psi(E)u_b(x) \,, \tag{1.6}$$

which is still a pure case.

The interaction will be described by a dipole interaction,

$$\hbar V = -\,e\,E\,x \quad . \tag{1.7}$$

During the interval $0 < t < \tau$ the wavefunction of the combined system will be

$$\begin{aligned} \Psi(E,x\,;\,t) &= \sum_n a_n(t)\, h_n(E)\, u_a(x)\; e^{-i(W_n + W_a)t} \\ &+ \sum_n b_n(t)\, h_n(E)\, u_b(x)\; e^{-i(W_n + W_b)t} \end{aligned} \tag{1.8}$$

The probability for the oscillator to be in the nth state at $t = \tau$ is given by the expression

$$\begin{aligned} P_n(\tau) &= \int |a_n(\tau)\, u_a(x)\, e^{-i\omega\tau} + b_n(\tau)\, u_b(x)\,|^2\, dx \\ &= |a_n(\tau)|^2 + |b_n(\tau)|^2 \,. \end{aligned} \tag{1.9}$$

The original pure case has been converted to a mixture, because we are not discussing the state of the reservoir.

To find the $a_n(t)$ and $b_n(t)$ in the interval $0 < t < \tau$, we must integrate the Schrödinger equation which for the resonance case considered can be reduced to a set of simultaneous algebraic equations :

$$i\dot{a}_{n-1} = V_n b_n \,, \quad i\dot{b}_n = V_n a_{n-1} \,, \tag{1.10}$$

where $n = 1, 2, \ldots,$ and

$$\hbar V_n = -\,e\langle n-1\,|E|\,n\rangle\langle a\,|x|\,b\rangle \quad . \tag{1.11}$$

The solution of equations (1.10) is

$$b_n(\tau) = b_n(0)\cos V_n\tau \,, \quad a_n(\tau) = -\,i\,b_{n+1}(0)\,\sin V_{n+1}\tau \,, \tag{1.12}$$

where $|b_n(0)|^2 = P_n(0)$. Hence we get for $P_n(\tau)$:

$$P_n(\tau) = P_n(0)\cos^2 V_n\tau + P_{n+1}(0)\sin^2 V_{n+1}\tau \,, \tag{1.13}$$

and

$$\delta P_n = P_n(\tau) - P_n(0) = P_{n+1}(0)\sin^2(V_{n+1}\tau) - P_n(0)\sin^2(V_n\tau) \,. \tag{1.14}$$

If we inject atoms at an average rate r', we can write a differential equation (coarse-graining ! ; again compare the Chapter on the Bloch equation) :

$$\frac{dP_n}{dt} = r' \left[P_{n+1}(t) \sin^2 (V_{n+1} \tau) - P_n(t) \sin^2 (V_n \tau) \right] . \tag{1.15}$$

We can easily generalize this equation by averaging with an exponential weight function $W(\tau) = \gamma e^{-\gamma\tau}$ over a distribution of interaction times (this corresponds to the case of a gas laser where active atoms are excited randomly and then radiatively decay exponentially). This means that the $\sin^2$ terms must be replaced by their averages

$$\langle \sin^2 V\tau \rangle = \frac{\frac{1}{2} V^2}{V^2 + \frac{1}{4}\gamma^2} . \tag{1.16}$$

We can further generalize by injecting atoms in state b (damping atoms) at a rate r' and atoms in state a (pumping atoms) at a rate r. In the case of a harmonic oscillator, V_n is proportional to $\sqrt{n}$ and we shall use the following notation

$$V_n^2 = nV^2 , \tag{1.17}$$

$$A = \tfrac{1}{2} r V^2 , \quad A' = \tfrac{1}{2} r' V^2 , \quad B = 2 r V^4/\gamma^2 , \quad B' = 2 r' V^4/\gamma^2 . \tag{1.18}$$

The final equation for P_n then becomes

$$\frac{dP_n}{dt} = - \frac{(n+1)A^2}{A + B(n+1)} P_n + \frac{nA^2}{A + Bn} P_{n-1} + \frac{(n+1)A'^2}{A' + B'(n+1)} P_{n+1} - \frac{nA'^2}{A' + B'n} P_n . \tag{1.19}$$

Although we have assumed V to be the same for the 'a' and the 'b' atoms, as we have introduced four variables, A, A', B, B', this restriction can be lifted as well.

Lamb discusses a number of special cases, and we shall consider two of them. First of all consider the case discussed when deriving equation (1.19) using equation (1.18), where $A/B = A'/B'$. If we put $C = A'$, we find a steady-state solution (corresponding to detailed-balance conditions) if

$$P_n = \frac{A}{C} P_{n-1} , \tag{1.20}$$

or

$$P_n = \left(\frac{A}{C}\right)^n P_0 , \tag{1.21}$$

with $P_0 = [1 - (A/C)]^{-1}$, if $A < C$. If we introduce a temperature T through the definition

$$\frac{A}{C} = \frac{r}{r'} = e^{-\beta\hbar\omega} , \quad \beta = \frac{1}{k_B T} , \tag{1.22}$$

equation (1.21) is the Planck distribution!

A simple quantum-mechanical theory of a laser is obtained by putting $B' = 0$. The damping atoms simulate the ohmic resistance of the cavity resonator, and the pumping atoms the effect of the population inversion produced, for instance, by an injected beam of state selected atoms. The solution now is

$$P_n = P_o \prod_{j=1}^{n} \frac{A/C}{1 + (jB/A)} . \tag{1.23}$$

If $A > C$, the P_n distribution will have a maximum at n_0 given by

$$\frac{A}{C} = 1 + \frac{B}{A} n_o . \tag{1.24}$$

corresponding to a laser above threshold.

8.2 Linear Response Theory

I now want to discuss a recent paper by van Kampen who presented 'The case against linear response theory'. The fact that — within a certain range of values — macroscopic relations often are linear must be understood on the basis of the microscopic equations of motion. In the usual approach to these problems one linearizes the microscopic equations of motion (see, for instance, the section 2.4 on ferromagnetic resonance) and van Kampen's case is that this way to obtain expressions for the phenomenological coefficients is a mathematical exercise (of doubtful validity) rather than a description of what actually happens.

Let us recall how linear response theory works. If a closed isolated system is subject to an external force $F(t)$, the Hamiltonian is

$$H_{op}(t) = H_{op} - F(t)\, Q_{op} . \tag{2.1}$$

To <u>first order</u> in F we have

$$\rho_{op}(t) = \rho_0 + i \int_{-\infty}^{t} F(t')\, dt' \left[Q_{op}(t'-t), \rho_0\right]_- , \tag{2.2}$$

with

$$\rho_0 = \exp(-\beta H_{op}) / \mathrm{Tr} \exp(-\beta H_{op}) , \tag{2.3}$$

and (we put $\hbar = 1$)

$$Q_{op}(t) = \exp(i H_{op} t)\, Q_{op} \exp(-i H_{op} t) . \tag{2.4}$$

Hence

$$\langle Q_{op}\rangle_t = \langle Q_{op}\rangle_o + i\int_{-\infty}^{t} F(t')\,dt'\,\langle [Q_{op}(t), Q_{op}(t')]_-\rangle_o \quad ,$$

or

$$\langle Q_{op}\rangle_t - \langle Q_{op}\rangle_o = \int_0^\infty \Gamma(\tau)\, F(t-\tau)\, d\tau \ , \tag{2.6}$$

with

$$\Gamma(\tau) = i\langle [Q_{op}, Q_{op}(-\tau)]_-\rangle_o \quad , \tag{2.7}$$

which means that the susceptibility $\chi(\omega)$ is frequency-dependent

$$\chi(\omega) = i\int_0^\infty \langle [Q_{op}, Q_{op}(-\tau)]_-\rangle_o\, e^{i\omega\tau}\, d\tau \ . \tag{2.8}$$

From (2.8) follow the usual properties of χ such as, for instance, $\chi(\omega) = -\chi^*(-\omega)$.

The point van Kampen made is that there is no reason to assume that linearity of the macroscopic response means that it is allowable to solve the equation for $\rho_{op}(t)$ to first order in F. It is unrealistic to assume that the microscopic behaviour should be linear in F; the linearity of macroscopic phenomena is a much more subtle effect and microscopic linearity is, in fact, incompatible with the basic ideas of the statistical mechanics of irreversible phenomena.

To estimate, for instance, for what fields the microscopic motions are linear in F we require that the distance travelled in time t by an electron, which is accelerated by an electric field, must be small compared to the distance d between electrons, or

$$\tfrac{1}{2} t^2 \frac{eF}{m} \ll d \ , \tag{2.9}$$

or, putting d as large as 0.1 cm ,

$$t^2 F \ll 10^{-18}\ \mathrm{V\ s^2/cm} \quad . \tag{2.10}$$

This should hold for times of the order of the duration of a macroscopic measurement, say $t = 1\,\mathrm{s}$, and hence

$$F \ll 10^{-18}\ \mathrm{V/cm}\ ! \tag{2.11}$$

Of course, once the linearity of the microscopic motion is shown to be wrong, the derivation of equation (2.6) falls and we have no longer a microscopic theory of the phenomenological coefficients.

Let us now consider the fundamental ideas of statistical mechanics. As stated at the beginning of these lecture notes, statistical mechanics is a way to explain the empirical fact that systems with huge numbers of degrees of freedom can be described on a macroscopic scale by a much smaller number of (macroscopic) variables which again obey deterministic equations of motion. In some sense the macroscopic variables are the slowly varying quantities of the system. Put differently, one can find time intervals Δt so small that the macroscopic variables are practically constant and so large that — with the macroscopic variables as constraints — the microscopic variables establish equilibrium: it is the randomization of the microscopic variables that enables one to eliminate them from the macroscopic picture.

In other words, statistical mechanics is concerned with the global aspects of the solutions of the equations of motion, that is, with asymptotic properties after sufficiently long times. Linear response theory, however, treats the equations of motion like the usual differential equations, that is, it looks at sufficiently small time intervals. Why has the lack of randomization in linear response theory not produced absurdities? The reason is that the linear approximation simulates the effect of randomization. Equilibrium is assumed at $t = -\infty$ and the equations of motion are solved to first order in F: the field therefore acts on the system as if it were in equilibrium. The trouble with linear response theory is that it ignores the previous history through linearization rather than forgets it through randomization.

Consider now — with van Kampen — a Galton board (Fig. 8.1). A large number of small balls are dropped from the hole at the top (A). The horizontal coordinate x and velocity v of a ball determine its microscopic state. Take, for the sake of simplicity, the vertical velocity of the balls to be a constant, which we shall take to be unity. Due to a spread in x, there will be a distribution of positions X where they hit the bottom, which one expects to be Gaussian with $\langle X \rangle = 0$. Apply now a horizontal

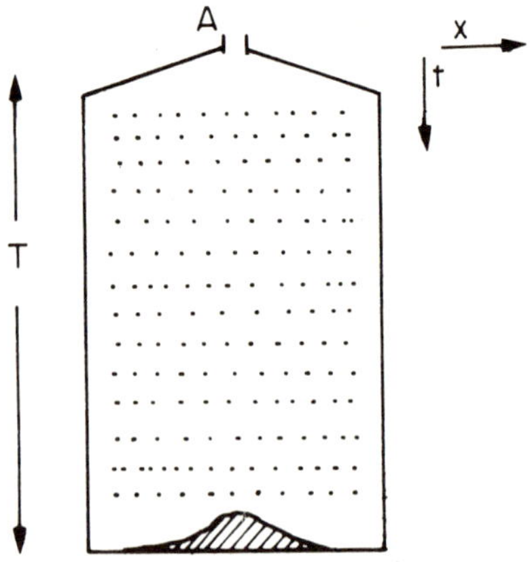

Fig. 8.1

force F (say, by tilting the board). Provided F is not too large $\langle X \rangle$ will shift by an amount which will be proportional to F. However, the individual paths will only show linear changes if the angle of tilting is small compared to a/T, where a is the sum of the diameters of the balls and of the pins and T the distance of A from the bottom.

Let us now analyze the process. Let $f(x,v,t)$ be the distribution at a height t (as the vertical velocity is unity, height and time are the same) below A. If $W(v|v')$ is the transition probability, and we introduce a collision integral

$$\left(\frac{\partial f}{\partial t}\right)_{coll} = \int \left[W(v|v')\ f(v') - W(v'|v)\ f(v)\right] dv' , \tag{2.12}$$

the kinetic equation for f will be of the form

$$\frac{\partial f}{\partial t} + v\frac{\partial f}{\partial x} + F\frac{\partial f}{\partial v} = \left(\frac{\partial f}{\partial t}\right)_{coll} . \tag{2.13}$$

Integrating equation (2.13) over x, we find for the velocity distribution function $g(v,t)$ the equation

$$\frac{\partial g}{\partial t} + F\frac{\partial g}{\partial v} = \left(\frac{\partial g}{\partial t}\right)_{coll} . \tag{2.14}$$

As $X = \langle v \rangle T$, we can measure the response to F through $\langle v \rangle$.

The basic ideas outlined earlier mean that apart possibly from the immediate vicinity of A, we are dealing with a steady state where $g(v,t)$ is independent of t so that

$$F\frac{\partial g}{\partial v} = \left(\frac{\partial g}{\partial t}\right)_{coll} , \tag{2.15}$$

and

$$\langle v \rangle = \int v\, g(v)\, dv \tag{2.16}$$

depends on F alone. Let $g_0(v)$ be the equilibrium distribution for the case $F = 0$. Writing

$$g(v\, ;\, F) = g_0(v) + F g_1(v) , \tag{2.17}$$

and linearizing the kinetic equation (that is, the macroscopic response!) we find

$$\frac{dg_0}{dv} = \left(\frac{dg_1}{dt}\right)_{coll} , \tag{2.18}$$

and once that equation has been solved, we find the phenomenological equation in the form

$$\langle v \rangle = F \int v g_1(v)\, dv \tag{2.19}$$

For the case when $W(v|v') = p(v)\, q(v')$ one can solve equation (2.18) explicitly, and one finds that

$$\int v\, g_1(v)\, dv = \frac{\displaystyle\int \frac{q - vq'}{q^2}\, dv}{\displaystyle\int p\, dv} \; . \tag{2.20}$$

Let us now draw attention to the difference between this kinetic approach and the linear response theory.

(1) Linearity here means expanding the distribution function rather than the individual paths and this is correct as long as $\langle v \rangle$ is small compared to the width of $g_0(v)$.

(2) As g is the steady-state solution, equation (2.18) does not contain t any longer, or, put differently, we solve equation (2.14) in the limit as $t \to \infty$, so that we have no problem with the adiabatic switching on of $F(t)$.

(3) We needed to use a randomization at each moment t (or after each time-interval Δt).

References

Note: As the foregoing text consists of lecture notes, I have not given detailed references as I would have done if it had been a textbook. However, it may be useful to give a few references for further reading.

CHAPTER 1

For a discussion of ensemble theory see :

ter Haar, D., Elements of Thermostatistics, Holt, Rinehart and Winston, New York, 1966.

For a discussion of the density matrix see:

ter Haar, D., Reports Progr. Phys., 24, 304 (1961).

The discussion of the occupation number representation is based on:

Schultz, T.D., Quantum Field Theory and the Many-Body Problem, Gordon and Breach, New York, 1964

de Boer, J., Studies Stat. Mech., 3, 215 (1965).

CHAPTER 2

For a general discussion of the temperature-dependent double-time Gree functions see:

Zubarev, D.N., Sov. Phys. - Uspekhi, 3, 320 (1960)

Bonch-Bruevich, V.L. and Tyablikov, S.V., Green Function Method in Statistical Mechanics, North-Holland, Amsterdam, 1962.

ter Haar, D., In Fluctuations, Relaxation and Resonance in Magnetic Systems, Oliver and Boyd, Edinburgh, p.119, 1962.

Section 2.3 is based upon

Bogolyubov, N.N. and Tyablikov, S.V., Sov. Phys. - Doklady, 4, 604 (1959).

See also:

Tahir-Kheli, R.A. and ter Haar, D., Phys. Rev., 127, 88 , 95 (1962).

and references given in the reference to section 2.7.

Section 2.4 see:

Tyablikov, S.V., Sov. Phys. - Solid State, 2, 332 , 1805 (1960).

Tyablikov, S.V. and Fu-Cho, Sov. Phys. - Solid State, 3, 102 (1961).

Section 2.5 see:

Fu-Cho, Sov. Phys..- Doklady, 5, 128, 321 (1960).

Lines, M.E., Phys. Rev., 131, 540 (1963).

Hewson, A.C. and ter Haar, D., Physica, 30, 890 (1964).

Hewson, A.C., ter Haar, D. and Lines, M.E., Phys. Rev., 137, A1465 (1965).

Section 2.6 is based upon:

Knapp Jr., R.H. and ter Haar, D., J. Stat. Phys., 1, 149 (1969).

Section 2.7 is based upon:

Lines, M.E., Phys. Rev., B3, 1749 (1971).

CHAPTER 3

This section is the text of an unpublished Clarendon Laboratory report by C.J. Pethick and D. ter Haar. It was part of C.J. Pethick's Oxford D.Phil. thesis. A short account was presented at the 1965 Novosibirsk Many-Body Conference and was published in the Proceedings of that conference. The appendix is based upon:

Bogolyubov, N.N, J. Phys. USSR, 11, 23 (1947).

CHAPTER 4

Section 4.1 is based upon:

Fujiwara, I., ter Haar, D. and Wergeland, H., J. Stat. Phys., 2, 329 (1970).

Section 4.2 is based upon:

't Hooft, A.H. and de Boer, J., Proc. Roy. Dutch Acad. Sc., B73, 443, 446 (1970).

CHAPTER 5

For a discussion of the equation of state of imperfect gases and of condensation phenomena, see, for instance:

ter Haar, D., Elements of Statistical Mechanics, Rinehart, New York, 1954.

Section 5.1 is based upon:

van Kampen, N.G., Phys. Rev., 135, A362 (1964).

see also

Lebowitz, J. and Penrose, O., J. Math. Phys., 7, 98 (1966).

Lieb, E., J. Math. Phys., 7, 1016 (1966).

Section 5.2 is based upon:

Kac, M., Uhlenbeck, G.E. and Hemmer, P.C., J. Math. Phys., 4, 216 (1963).

see also

Hemmer, P.C., Kac, M. and Uhlenbeck, G.E., J. Math. Phys., 4, 229 (1963).

Uhlenbeck, G.E., Hemmer, P.C. and Kac, M., J. Math. Phys., 5, 60 (1964).

CHAPTER 6

This section is based upon:

ter Haar, D., Am. J. Phys., 34, 1164 (1966);

see also

Stenholm, S. and ter Haar, D., Physica, 32, 136 (1966).

CHAPTER 7

For an introduction to the statistical mechanics of stellar systems, see

ter Haar, D., J. Phys. Soc. Japan, 26 S, 25 (1969).

Section 7.2 is based upon:

Lynden-Bell, D., Mon. Not. Roy. Astron. Soc., 136, 101 (1967).

Section 7.3 is based upon:

ter Haar, D., Men of Physics: L.D. Landau, vol.II, Pergamon Press, Oxford, 1969. (Introduction).

Lynden-Bell, D., Mon. Not. Roy. Astron. Soc., 124, 95 (1962).

Section 7.4

Lynden-Bell, D., Mon. Not. Roy. Astron. Soc., 136, 101 (1967).

Saslaw, W.C., Mon. Not. Roy. Astron. Soc., 141, 77 (1968).

Saslaw, W.C., Mon. Not. Roy. Astron. Soc., 143, 437 (1969).

Saslaw, W.C., Mon. Not. Roy. Astron. Soc., 147, 253 (1970).

Saslaw, W.C., Mon. Not. Roy. Astron. Soc., 150, 299 (1970).

Lynden-Bell, D. and Wood, R., Mon. Not. Roy. Astron. Soc., 138, 495 (1968).

Section 7 5 is based upon:

Gilbert, I.H., Astrophys. J., 152, 1043 (1968).

see also

Gilbert, I.H., Astrophys. J., 159, 239 (1970).

CHAPTER 8

Section 8.1 is based upon:

Lamb, W.E. Jr., In The Physicist's Conception of Nature, (J. Mehra editor), Reidel, Dordrecht, 1973.

Section 8.2 is based upon:

van Kampen, N.G., Physica Norvegica, 5, 279 (1971).

Index

OTHER TITLES IN THE SERIES IN NATURAL PHILOSOPHY

Vol. 1. Davydov—Quantum Mechanics
Vol. 2. Fokker—Time and Space, Weight and Inertia
Vol. 3. Kaplan—Interstellar Gas Dynamics
Vol. 4. Abrikosov, Gor'kov and Dzyaloshinskii—Quantum Field Theoretical Methods in Statistical Physics
Vol. 5. Okun'—Weak Interaction of Elementary Particles
Vol. 6. Shklovskii—Physics of the Solar Corona
Vol. 7. Akhiezer *et al.*—Collective Oscillations in a Plasma
Vol. 8. Kirzhnits—Field Theoretical Methods in Many-body Systems
Vol. 9. Klimontovich—The Statistical Theory of Non-equilibrium Processes in a Plasma
Vol. 10. Kurth—Introduction to Stellar Statistics
Vol. 11. Chalmers—Atmospheric Electricity (2nd Edition)
Vol. 12. Renner—Current Algebras and their Applications
Vol. 13. Fain and Khanin—Quantum Electronics, Volume 1—Basic Theory
Vol. 14. Fain and Khanin—Quantum Electronics, Volume 2—Maser Amplifiers and Oscillators
Vol. 15. March—Liquid Metals
Vol. 16. Hori—Spectral Properties of Disordered Chains and Lattices
Vol. 17. Saint James, Thomas and Sarma—Type II Superconductivity
Vol. 18. Margenau and Kestner—Theory of Intermolecular Forces (2nd Edition)
Vol. 19. Jancel—Foundations of Classical and Quantum Statistical Mechanics
Vol. 20. Takahashi—An Introduction to Field Quantization
Vol. 21. Yvon—Correlations and Entropy in Classical Statistical Mechanics
Vol. 22. Penrose—Foundations of Statistical Mechanics
Vol. 23. Visconti—Quantum Field Theory, Volume 1
Vol. 24. Furth—Fundamental Principles of Theoretical Physics
Vol. 25. Zheleznyakov—Radioemission of the Sun and Planets
Vol. 26. Grindlay—An Introduction to the Phenomenological Theory of Ferroelectricity
Vol. 27. Unger—Introduction to Quantum Electronics
Vol. 28. Koga—Introduction to Kinetic Theory: Stochastic Processes in Gaseous Systems
Vol. 29. Galasiewicz—Superconductivity and Quantum Fluids
Vol. 30. Constantinescu and Magyari—Problems in Quantum Mechanics
Vol. 31. Kotkin and Serbo—Collection of Problems in Classical Mechanics
Vol. 32. Panchev—Random Functions and Turbulence
Vol. 33. Talpe—Theory of Experiments in Paramagnetic Resonance
Vol. 34. ter Haar—Elements of Hamiltonian Mechanics (2nd Edition)
Vol. 35. Clarke and Grainger—Polarized Light and Optical Measurement
Vol. 36. Haug—Theoretical Solid State Physics, Volume 1
Vol. 37. Jordan and Beer—The Expanding Earth
Vol. 38. Todorov—Analytical Properties of Feynman Diagrams in Quantum Field Theory
Vol. 39. Sitenko—Lectures in Scattering Theory
Vol. 40. Sobel'man—Introduction to the Theory of Atomic Spectra
Vol. 41. Armstrong and Nicholls—Emission, Absorption and Transfer of Radiation in Heated Atmospheres
Vol. 42. Brush—Kinetic Theory, Volume 3
Vol. 43. Bogolyubov—A Method for Studying Model Hamiltonians
Vol. 44. Tsytovich—An Introduction to the Theory of Plasma Turbulence
Vol. 45. Pathria—Statistical Mechanics
Vol. 46. Haug—Theoretical Solid State Physics, Volume 2
Vol. 47. Nieto—The Titius-Bode Law of Planetary Distances: Its History and Theory
Vol. 48. Wagner—Introduction to the Theory of Magnetism
Vol. 49. Irvine—Nuclear Structure Theory
Vol. 50. Strohmeier—Variable Stars
Vol. 51. Batten—Binary and Multiple Systems of Stars
Vol. 52. Rousseau and Mathieu—Problems in Optics
Vol. 53. Bowler—Nuclear Physics
Vol. 54. Pomraning—The Equations of Radiation Hydrodynamics
Vol. 55. Belinfante—A Survey of Hidden Variables Theories
Vol. 56. Scheibe—The Logical Analysis of Quantum Mechanics
Vol. 57. Robinson—Macroscopic Electromagnetism
Vol. 58. Gombas and Kisdi—Wave Mechanics and Its Applications
Vol. 59. Kaplan and Tsytovich—Plasma Astrophysics
Vol. 60. Kovacs and Zsoldos—Dislocations and Plastic Deformation
Vol. 61. Auvray and Fourrier—Problems in Electronics
Vol. 62. Mathieu—Optics
Vol. 63. Atwater—Introduction to General Relativity
Vol. 64. Muller—Quantum Mechanics: A Physical World Picture
Vol. 65. Bilenky—Introduction to Feynman Diagrams
Vol. 66. Vodar and Romand—Some Aspects of Vacuum Ultraviolet Radiation Physics
Vol. 67. Willett—Gas Lasers: Population Inversion Mechanisms
Vol. 68. Akhiezer *et al.*—Plasma Electrodynamics, Volume 1—Linear Theory
Vol. 69. Glasby—The Nebular Variables
Vol. 70. Bialynicki-Birula—Quantum Electrodynamics
Vol. 71. Karpman—Non-linear Waves in Dispersive Media
Vol. 72. Cracknell—Magnetism in Crystalline Materials
Vol. 73. Pathria—The Theory of Relativity
Vol. 74. Sitenko and Tartakovskii—Lectures on the Theory of the Nucleus
Vol. 75. Belinfante—Measurement and Time Reversal in Objective Quantum Theory
Vol. 76. Sobolev—Light Scattering in Planetary Atmospheres
Vol. 77. Novakovic—The Pseudo-spin Method in Magnetism and Ferroelectricity
Vol. 78. Novozhilov—Introduction to Elementary Particle Theory
Vol. 79. Busch and Schade—Lectures on Solid State Physics
Vol. 80. Akhiezer *et al.*—Plasma Electrodynamics, Volume 2
Vol. 81. Soloviev—Theory of Complex Nuclei
Vol. 82. Taylor—Mechanics: Classical and Quantum
Vol. 83. Srinivasan and Parthasathy—Some Statistical Applications in X-Ray Crystallography
Vol. 84. Rogers—A Short Course in Cloud Physics
Vol. 85. Ainsworth—Mechanisms of Speech Recognition
Vol. 86. Bowler—Gravitation and Relativity
Vol. 87. Davydov—Quantum Mechanics (2nd Edition)
Vol. 88. Weiland and Wilhelmsson—Coherent Non-Linear Interaction of Waves in Plasmas
Vol. 89. Pacholczyk—Radio Galaxies
Vol. 90. Elgarøy—Solar Noise Storms
Vol. 91. Heine—Group Theory in Quantum Mechanics